Climate Policy Integration into EU Energy Policy

Climate change is a cross-cutting, long-term, global problem that presents policymakers with many challenges in their efforts to respond to the issue. Integrating climate policy objectives into the elaboration and agreement of policy measures in other sectors represents one promising method for ensuring coherent policies that respond adequately to the climate change challenge.

This book explores the integration of long-term climate policy objectives into EU energy policy. It engages in in-depth empirical analysis on the integration of climate policy objectives into renewable energy policy; energy performance of buildings; and policies in support of natural gas importing infrastructure. The book describes insufficient levels of climate policy integration across these areas to achieve the long-term policy goals. A conceptual framework to find reasons for insufficient integration levels is developed and applied.

This book is a valuable resource for students, researchers, academics and policymakers interested in environmental, climate change and energy policy development in the EU, particularly from the perspective of long-term policy challenges. The book adds to scholarly literature on policy integration and EU integration, and contributes to new and developing research about EU decarbonisation.

Claire Dupont is a post-doctoral researcher at the Institute for European Studies and the Political Science department of the Vrije Universiteit Brussel (VUB). She works within the VUB's Strategic Research Programme on 'Evaluating Democratic Governance in Europe (EDGE)'.

Routledge Studies in Energy Policy

Our Energy Future
Socioeconomic implications and policy options for rural America
Edited by Don E. Albrecht

Energy Security and Natural Gas Markets in Europe
Lessons from the EU and the USA
Tim Boersma

International Energy Policy
The emerging contours
Edited by Lakshman Guruswamy

Climate Policy Integration into EU Energy Policy
Progress and prospects
Claire Dupont

For further details please visit the series page on the Routledge website:
www.routledge.com/books/series/RSIEP/

Climate Policy Integration into EU Energy Policy

Progress and prospects

Claire Dupont

First published 2016
by Routledge
2 Park Square, Milton Park, Abingdon, Oxon OX14 4RN

and by Routledge
711 Third Avenue, New York, NY 10017

Routledge is an imprint of the Taylor & Francis Group, an informa business

© 2016 Claire Dupont

The right of Claire Dupont to be identified as author of this work has been asserted by him/her in accordance with sections 77 and 78 of the Copyright, Designs and Patents Act 1988.

All rights reserved. No part of this book may be reprinted or reproduced or utilised in any form or by any electronic, mechanical, or other means, now known or hereafter invented, including photocopying and recording, or in any information storage or retrieval system, without permission in writing from the publishers.

Trademark notice: Product or corporate names may be trademarks or registered trademarks, and are used only for identification and explanation without intent to infringe.

British Library Cataloguing-in-Publication Data
A catalogue record for this book is available from the British Library

Library of Congress Cataloging-in-Publication Data
Dupont, Claire.
Climate policy integration into EU energy policy : progress and prospects / by Claire Dupont.
pages cm
ISBN 978-1-138-80646-7 (hb) — ISBN 978-1-315-75166-5 (e-book) 1. Energy policy—Environmental aspects—European Union countries. 2. Climatic changes—Government policy—European Union countries. 3. Environmental policy—European Union countries. I. Title.
HD9502.E852D86 2016
333.79094—dc23
2015016082

ISBN: 978-1-138-80646-7 (hbk)
ISBN: 978-1-315-75166-5 (ebk)

Typeset in Goudy
by FiSH Books Ltd, Enfield

Printed and bound by CPI Group (UK) Ltd, Croydon, CR0 4YY

For Nico

Contents

List of figures	viii
List of tables	x
Acknowledgements	xi
List of acronyms and abbreviations	xii
1 Introduction	1
2 From environmental to climate policy integration	30
3 EU policy on renewable energy	61
4 EU policy on the energy performance of buildings	95
5 EU policy on natural gas import infrastructure	127
6 Explaining climate policy integration: policy, politics, context and process	152
7 Conclusions	174
List of interviews	188
Index	189

Figures

3.1 Percentage share of renewable energy in final energy consumption in the EU28 — 62
3.2 Expected very high levels of CPI from 2000–2010, compared to BAU scenarios from 2006 and 2011 — 66
3.3 Policy output of the 2001 RES-E Directive compared to BAU and to very high levels of CPI — 79
3.4 CPI in the policy output of the 2001 RES-E Directive, assuming lower advances in RE share between 2000 and 2010 than between 2010 and 2050 — 80
3.5 CPI in the policy output of the 2009 RE Directive compared to BAU scenarios and the very high CPI trajectory from 2005 levels — 87
3.6 CPI trajectory for the RE share to 2050, with lower expectations to 2020 — 88
4.1 Development of energy intensity of the economy of the EU28 between 2002 and 2013, measured in kilograms of oil equivalent per €1000 — 96
4.2 Final energy consumption in the EU28 from 2000 to 2013, measured in mega tonnes of oil equivalent — 96
4.3 Very high CPI trajectory for energy consumption in buildings in EU28, from 2000 to 2050, compared to the BAU trajectory — 102
4.4 CPI in the policy output of the 2002 EPBD, with measures expected to avoid the consumption of 55 Mtoe of energy in buildings in the EU between 2000 and 2010 — 112
4.5 EPBD measures compared to actual consumption of energy in buildings, 2000–2010 — 113
4.6 CPI in the policy output of the 2010 EPBD recast, with measures expected to avoid between 60 and 80 Mtoe of energy consumption in buildings in 2020 — 120
4.7 Updated BAU scenario from 2011, which includes expected EPBD impact on the energy consumption in buildings to 2020, compared to the 2006 BAU trajectory and CPI ranges to 2020 — 121

5.1 Range of very high CPI gas consumption trajectories in the EU to 2050, compared to the BAU and decarbonisation scenarios of the Commission's Energy Roadmap   134
5.2 Very high levels of CPI to 2050, with continued natural gas consumption to 2020 before consumption begins to reduce in line with decarbonisation scenarios   134

Tables

1.1	Some of the main policy developments in EU energy and climate policy, 1950s–1980s	15
1.2	Some of the main policy developments in EU energy and climate policy in the 1990s	17
1.3	Some of the main policy developments in EU energy and climate policy in the 2000s	20
2.1	Measuring the level of CPI in the policy process	39
2.2	Scale establishing the extent of CPI in the policy output	40
2.3	The nature of functional interrelations and their potential effect on CPI	51
2.4	Measuring political commitment to, first, combating climate change, and, second, advancing CPI	52
2.5	Summary of explanatory framework	55
3.1	Percentage share of RE in final energy consumption in the EU28 and in each member state in 2004 and 2013, compared to 2020 target	63
4.1	Some of the main pieces of EU legislation promoting energy efficiency	98
5.1	EU gas import capacity, circa 2012 to 2022, taking account of projects under construction, but excluding proposed projects	129
6.1	Summary of levels of CPI in the policy process and output of the case studies	153
6.2	The nature of functional interrelations and the expected effect on levels of CPI	157
6.3	Overarching political commitment to combating climate change and to advancing CPI in the cases	161
6.4	The potential of the internal institutional context and external policy context to advance CPI in the cases	166
6.5	Summary of the explanatory variables for levels of CPI in the case studies	169

Acknowledgements

This book results from the work carried out during and following my PhD research at the Institute for European Studies (IES) at the Vrije Universiteit Brussel (VUB). The book could not have been completed without the financial support provided by the IES and the VUB's Strategic Research Programme 'Evaluating Democratic Governance in Europe (EDGE)'. The EDGE financial support allowed me to work on the final manuscript between 2014 and 2015, and provided me with the freedom to dedicate time to the book. I am also indebted to the Academy of Finland for providing support, under the Ecoherence project, for a research visit at the University of Turku in January and February 2015 that helped me finalise the book.

A great many individuals also contributed to ensuring this book improved after each draft. I am particularly grateful for the support and constant feedback of Sebastian Oberthür, who has followed my progress since the beginning. Harri Kalimo and Ingmar von Homeyer repeatedly provided me with challenging and helpful feedback on early drafts, which improved the final product immeasurably. Andrea Lenschow, Camilla Adelle and Irina Tanasescu gave very helpful comments on later versions of the work that allowed me to streamline the final book.

I would also like to thank the staff of the IES for their constant hard work in supporting my research. In particular, I am grateful to the members of the Environment and Sustainable Development research cluster at the IES and to members of the secretariat who are always willing to help.

This book could not have been completed without the support of friends and family. It is dedicated to Nico, who has constantly encouraged me in my work.

Brussels
March 2015

Acronyms and abbreviations

ALDE	Alliance of Liberals and Democrats for Europe
BAU	Business-as-usual
bcm	Billion cubic metres
BP	British Petroleum
BPIE	Buildings Performance Institute Europe
CAIT	Climate analysis indicator tool
CCS	Carbon capture and storage
CEF	Connecting Europe Facility
CH_4	Methane
CO_2	Carbon dioxide
COP	Conference of the Parties
COR	Committee of the Regions
COREPER	Committee of Permanent Representatives
CPI	Climate policy integration
DG	Directorate General
EAP	Environmental action programme
ECCP	European climate change programme
ECF	European Climate Foundation
EEA	European Environment Agency
EEPR	European energy programme for recovery
EESC	European Economic and Social Committee
ENVI	Environment committee in the European Parliament
EPB	Energy performance of buildings
EPBD	Energy performance of buildings Directive
EPI	Environmental policy integration
EPP	European People's Party
EREC	European Renewable Energy Council
ETS	Emissions trading scheme
EU	European Union
Euratom	European Atomic Energy Community
EWEA	European Wind Energy Association
GHG	Greenhouse gas
HFC	Hydrofluorocarbon

IEA	International Energy Agency
IPCC	Intergovernmental Panel on Climate Change
ITGI	Italy-Greece-Turkey interconnector
ITRE	Energy committee in the European Parliament
KP	Kyoto Protocol
LNG	Liquefied natural gas
MEP	Member of the European Parliament
Mtoe	Mega tonnes of oil equivalent
MW	Megawatt
N_2O	Nitrous oxide
NGO	Non-governmental organisation
OECD	Organisation for Economic Cooperation and Development
PFC	Perfluorocarbon
QMV	Qualified majority voting
RE	Renewable energy
RES-E	Renewable energy sources of electricity
S&D	Group of the Progressive Alliance of Socialists and Democrats
SEA	Single European Act
SF_6	Sulphur hexafluoride
TAP	Trans-Adriatic pipeline
tcm	Trillion cubic metres
TEC	Treaty establishing the European Community
TEN-E	Trans-European networks for energy
TFEU	Treaty on the functioning of the European Union
UNEP	United Nations Environment Programme
UNFCCC	United Nations Framework Convention on Climate Change
WMO	World Meteorological Organisation
WWF	World Wide Fund for Nature

1 Introduction

Climate change is often considered a 'wicked' problem (Jordan, Huitema, van Asselt, Rayner and Berkhout, 2010: 4–5). It is a cross-cutting, long-term, global problem that presents policymakers with many challenges in their efforts to respond to the issue. The European Union (EU) has developed a leadership stance on the climate issue over time, and aims to advance global agreement through diplomacy and credible leadership by example (Wurzel and Connelly, 2011b). Ensuring that its domestic policies are sufficiently ambitious to demonstrate to the global community the potential and importance of advancing policies to combat climate change is one of the EU's leadership strategies (Oberthür and Roche Kelly, 2008).

Climate change is more than an environmental issue. With the effects of climate change expected to increase pressure on our food and energy systems, combating climate change (and adapting to its impacts) is a matter of survival. The fifth assessment report of the Intergovernmental Panel on Climate Change (IPCC) highlighted that it is unequivocal that climate change is caused by human activity – and most importantly by the burning of fossil fuels (such as coal, oil and gas), which emits dangerous greenhouse gases (GHGs). Over time, these GHGs accumulate in the earth's atmosphere, so that less heat escapes into space and global temperatures increase (the 'greenhouse effect') (IPCC, 2007, 2013). But responding to climate change is a challenge for policymakers. It is a global crisis that cuts across frontiers, affects developed and developing countries, and requires action and policy measures across a wide variety of policy fields. Policy sectors that affect climate change, or that will be affected by climate change, include agriculture, biodiversity, energy, fisheries, health, industry, migration, transport, waste management, water, among others. Effectively combating climate change means ensuring that climate change is 'integrated' or 'mainstreamed' into the policy process and policy output of each of these policy sectors. As a 'wicked problem', climate change belongs to a set of problems that 'challenge established social values and institutional frameworks, defy analysis, and have no obvious solutions' (Jordan *et al.*, 2010: 4–5). While such a framing may seem defeatist, it is certainly true that, despite growing scientific evidence, humanity has nonetheless continued its fossil fuel-consuming development path that upsets the precarious balance in the climate system (IEA, 2011, 2013).

Despite these challenges, the EU has developed several policy measures over the years in its effort to exert international leadership in the fight against climate change. Since the early 1990s, the EU has expressed a desire to lead the international community in solving this issue and it has increasingly added credibility to its rhetoric by agreeing internal policy measures to reduce its own emissions of GHGs (Gupta and Grubb, 2000; Oberthür and Roche Kelly, 2008; Wurzel and Connelly, 2011b). EU citizens generally view climate change as an important issue, and actions to combat climate change in the EU have received popular support (Eurobarometer, 2008, 2011, 2014). These realities mean that climate change has enjoyed a higher status than many other environmental issues in the EU. Given the complex reality of the climate issue, however, it is clear that environmental policies alone are insufficient to address the problem. With many other sectors *causing* climate change, responses need to be developed in these sectors in particular. Climate policy integration (CPI), then, represents one possible policy strategy that can be deployed to respond to the challenge of making policy to combat climate change.

But what is meant by *climate policy integration*? In some respects, climate policy integration is an *example* of *environmental* policy integration (EPI). Article 11 of the Treaty on the Functioning of the European Union (TFEU) states: 'Environmental protection requirements must be integrated into the definition and implementation of the Union's policies and activities, in particular with a view to promoting sustainable development', providing a legal obligation on the EU to EPI, but what precisely EPI is in practice has also long been discussed. Whether EPI is considered a policy or legal principle (Nollkamper, 2002), an objective for improving the efficiency of the policy process or a requirement for policies to improve the state of the environment (Persson, 2004), policymakers and scholars have not managed to agree. At the very minimum, 'integration' implies that one policy's objectives become part of another policy's development (Briassoulis, 2005).

The focus of this book is on the EU's internal energy policies and on whether they sufficiently integrate climate policy objectives. I describe and analyse the level of CPI in EU energy policy between 2000 and 2010, with the aim of explaining why varying levels of CPI within the EU's energy policy exist. This research falls firmly within, and adds to, academic literature on environmental and climate policy integration (Adelle and Russel, 2013; Jordan and Lenschow, 2010; Lafferty and Hovden, 2003; Lenschow, 2002; Nilsson and Eckerberg, 2007). The book presents the results of an in-depth case study analysis of EU energy policies, and a testing ground for operationalising and explaining CPI. The results provide valuable reflection, not only for academic research on EPI and CPI, but also for policymakers engaged in the climate and energy policy sectors.

In this introduction, I first describe the research puzzle and main research question guiding the book. Next I present the research set-up and design. I then describe the EU's policymaking processes and some of the key moments in the historical development of climate and energy policy in the EU. Finally, I present the overall structure of the book.

Research puzzle and design

With the adoption of the 2009 'integrated' package of policy measures on climate and energy, it seemed not only that the EU had finally followed through on its promises of climate leadership, but that it had also practised EPI (Adelle, Pallemaerts and Chiavari, 2009; Adelle, Russel and Pallemaerts, 2012). Nevertheless, this development also raised several questions that form the central motivations for the research described in this book:

- Just how 'integrated' are climate policies into EU energy policies anyway?
- Has the level of CPI in EU energy policy changed over time?
- Is CPI uniform across energy policies?
- How could potential variances in the levels of CPI be explained?

As a result of reflections on these questions, the main research question guiding the book can be formulated as follows:

> What is the extent of climate policy integration into the EU's energy policy, and why?

Thus, 'the extent of climate policy integration' is the dependent variable under investigation within the broad case of 'EU energy policy'. Independent variables are sought to explain the dependent variable (see Chapter 2).

The focus on energy policy is justified due to the energy sector being the biggest contributor of GHG emissions in the EU. Whether it is through the production, transmission or consumption of energy, GHG emissions from energy overall are estimated to be at the source of about 80 per cent of total GHG emissions in the EU (EEA, 2010: 31). Mitigating (or combating) climate change requires the reduction of GHG emissions as soon as possible, and by 2050 to a level 80 to 95 per cent lower than the level of GHG emissions in the EU in 1990. Any attempt to mitigate climate change, therefore, must consider policy action in the energy sector. This 80 to 95 per cent reduction in GHG emissions in the EU is a political commitment agreed by the European Council in October 2009 (European Council, 2009). It represents a translation of scientific estimates of the effort required to avoid dangerous anthropogenic interference with the climate system, and to ensure that global temperature increase does not exceed two degrees Celsius (European Commission, 2007b). This 2050 target to reduce GHG emissions by 80 to 95 per cent in the EU is thus both a scientific target and a political commitment. The EU is already on its way to reducing its GHG emissions compared to 1990 levels. By 2013, the EU had reduced its emissions by about 19 per cent (EEA, 2014). This is already an achievement and a policy success, but the remaining effort required for 2050 is considerable. It would not be possible to meet the 2050 objective without adopting policy measures in the energy sector that ensure the reduction of GHG emissions.

4   Introduction

In 2011, the Commission released its 'Roadmap for moving to a competitive low carbon economy in 2050' (European Commission, 2011b), which recognised that the greatest reductions in GHG emissions will need to occur in sectors that produce and consume energy. Certain sectors of the economy will face greater obstacles in reducing their GHG emissions (agriculture, for example). This communication was followed by a detailed roadmap for the energy sector (European Commission, 2011a), which outlines a number of scenarios for the energy sector's contribution to the EU's effort to reduce GHG emissions. This effort amounts to at least an 85 per cent reduction in GHG emissions from the energy sector (ibid.: 2). It is thus clear that aspirations exist within the EU to ensure that climate policy objectives are well integrated into EU energy policy (as also demonstrated by the above-mentioned 2009 climate and energy package of policy measures). The energy sector is a crucial sector for combating climate change. Is CPI in the energy sector sufficient to meet the long-term climate policy objectives? The energy sector should provide rich empirical data for understanding and explaining levels of CPI.

The research presented here is of primarily a *qualitative* nature, although certain parts contain some *quantitative* elements (for example, in terms of analysing statistics on GHG emissions, energy consumption and production or renewable energy generation). The main research strategy deployed involves analysing a limited number of *case studies* of EU energy policy to establish the levels of CPI in their policy processes and outputs (George and Bennett, 2004; Gerring, 2007; King, Keohane and Verba, 1994; Swanborn, 2010; Yin, 2009). Analysis in the case studies follows a *process tracing* strategy (Bennett, 2008; Checkel, 2008), including several data-collection and analysis techniques: *document analysis, literature review*, and complementary semi-structured *interviews* (Hopf, 2004; Kvale and Brinkman, 2009; Rathbun, 2008).

To answer the research question, I selected a number of cases of EU energy policy to examine in some detail. The entire EU energy portfolio is too large for a qualitative case study, so the case selection process followed a number of steps and criteria. The universe of cases from which I chose included all EU energy policies. After an initial survey of this universe of cases, I found that they could be categorised into three main types (linked to the categories of EU competence on energy policy in Article 194.1, TFEU):

1   EU internal policies on the production and transmission of energy;
2   EU internal policies on the consumption of energy;
3   EU external policies on ensuring continued supplies of energy to the EU.

The first category of energy policies includes policy measures on, for example, renewable energy generation, the internal market for energy, and on coal and carbon capture and storage (CCS) technologies. The second category of energy policies includes, for example, policy measures to reduce energy consumption by improving the energy-use of products, energy labelling and the energy performance of buildings, among others. Finally, the third category of policies is related to external relations. This category deals specifically with ensuring the security of energy supplies, for which only limited competence exists at the EU level. Bilateral negotiations for supplies of energy are often in the hands of member states, but the EU itself is also involved in energy dialogues with key energy partners, such as Russia (Hadfield, 2008; interview 9). Other policy measures at EU level in this third category include, for example, policies on stocks of natural gas and oil, policies in response to energy supply emergencies and policies on infrastructure measures for imports of energy sources.

I selected three cases for analysis based on expected variation on the variables of interest (Gerring, 2007, 2008: 668; Yin, 2009). Thus, the case studies chosen vary in terms of the *dependent variable* (the level of CPI). Some variation on the *independent variables* (namely, functional interrelations; political commitment; institutional and policy context; and the process dimension; see Chapter 2) is thus also expected, where the variation of the independent variables explains the variation on the dependent variable (King *et al.*, 1994: 91–93).

Nevertheless, the case studies are bounded within a specific universe (Klotz, 2008; Rohlfing, 2012: 24–28). First, the cases chosen are *geographically* bounded, in that I focus on EU energy policies that have effect internally to the EU. Second, the cases are *institutionally* bounded, in that the cases chosen were all agreed under the ordinary legislative procedure (previously the co-decision procedure) of the EU (with the Commission proposing legislation and the Parliament and Council co-deciding, see below). Third, the case studies are *substantively* bounded as they are all drawn from the energy policy domain. Fourth, the cases are *temporally* bounded, with policy developments occurring between 2000 and 2010. To see potential developments over time, I chose cases that had measures adopted in the early and late years of this decade. Additionally, and within these boundaries, I aimed for a selection of cases that drew from as many types of EU energy policy as possible generally, while still ensuring the number of cases was limited enough for manageable and in-depth research. I therefore chose one case from each of the three categories mentioned above.

The first case study chosen, described in Chapter 3, is on EU renewable energy (RE) policy. Two policy instruments developed between 2000 and 2010 form the centre of analysis for this case study: the 2001 Directive 2001/77/EC on the promotion of electricity produced from renewable energy sources (RES-E Directive) and the 2009 Directive 2009/28/EC on the promotion of the use of energy from renewable sources (RE Directive). The RE case is expected to demonstrate *high* levels of CPI in the policy process and output in the latter half of the 2000s, especially given the synergistic interrelations between climate policy objectives to reduce emissions of GHGs (especially from fossil fuels) and the objectives of RE policy (to increase the share of RE in the EU). This is expected to develop over time from more *medium* levels in the early 2000s.

The second case study, discussed in Chapter 4, comes from the second category of EU energy policy measures and is on the energy performance of buildings (EPB). Again, there are two policy instruments under focus in this case: the 2002

Directive 2002/91/EC on the energy performance of buildings and its 2010 recast Directive 2010/31/EU. This case is expected to produce different results on the dependent variable – with more *medium* levels of CPI expected overall in the policy process and output of the 2010 Directive, evolving from *low to medium* levels in the 2002 Directive. Given the synergistic interrelations between reducing energy consumption and reducing GHG emissions, the *medium* levels expected are related to issues of political commitment to agree on ambitious action to improve the energy performance of buildings. For both the 2002 and 2010 Directives, non-binding energy efficiency targets were in place, which hints at lower commitment than in RE policy to achieving the goals. This policy sector has also historically faced problems of implementation and ambition (Boasson and Dupont, 2015; Boasson and Wettestad, 2013).

Finally, the third case study, linked to the third category of EU energy policy measures listed above, relates to policies that promote the development of importing natural gas infrastructure (see Chapter 5). There are again two policy instruments that are examined in this case. The first is the trans-European networks for energy guidelines (TEN-E). These guidelines were revised in 2003 and in 2006 (Decisions 1229/2003/EC and 1364/2006/EC laying down guidelines for trans-European energy networks). The second policy instrument is the 2009 Regulation on the European Energy Programme for Recovery (EEPR, Regulation (EC) No. 663/2009). These policy instruments support energy infrastructure projects in both natural gas and electricity (and in the case of the EEPR Regulation, also offshore wind and CCS projects), both internally and externally to the EU. The analysis focuses on the external natural gas infrastructure projects that are supported by these policy instruments. In this case, *low* levels of CPI are expected over the time 2000–2010, as the promotion of further gas pipelines interrelates in a conflictual manner with long-term climate policy objectives, by potentially locking energy infrastructure into a fossil fuel pathway. The continued promotion of these policy measures runs counter to objectives to advance CPI.

The case study selection aims to present variation on the dependent variable within the gamut of potential EU energy policies by choosing one case from each of the three categories in the loose categorisation mentioned above. However, given the limited number of case studies, it may be difficult to generalise the results to overall EU energy policy, and further empirical research is likely to be required to test these results against cases of CPI in other policy sectors. As mentioned, the case studies are analysed using *process tracing* techniques, including: *document analysis, literature review* and semi-structured *interviews* with individuals involved in policy development.

Process tracing involves 'looking at evidence within an individual case' (Bennett, 2008: 704). It means providing an historical narrative of a case that produces multiple observations or pieces of evidence that could outline a causal chain linking the independent and dependent variables (Checkel, 2008: 115). The data used in process tracing is usually qualitative in nature, but in this book I also present some data from quantitative sources. The sources of data include academic literature and texts, databases of climate and energy statistics, policy documents, policy reports, position papers, media reports and interview data. Interviews served a complementary data collection function (Hopf, 2004: 203), rather than a primary data source. The interviews were carried out for the purpose of confirming/challenging information on the level of CPI in the policy process and for seeking opinions of policymakers and stakeholders. The interviews therefore were *semi-structured* with some guiding questions prepared in advance that allowed enough flexibility for a dialogue to develop (Hermanns, 2004; Hopf, 2004; Rathbun, 2008). The interviews were carried out on the basis of anonymity, and respondents spoke of their personal experience within the context of their organisations.

EU policymaking in climate and energy: an historical overview

Before embarking on a study of climate policy integration into the EU's energy policy, it is necessary to situate the study in the broader context of EU policymaking in general and in EU climate and energy policy development over time. In this section, I describe the EU's main policymaking process (the ordinary legislative procedure, previously known as the co-decision procedure). This information is necessary for understanding the analysis of CPI in the policy processes in the cases. Next, I describe the main historical developments in EU climate and energy policy, to provide a background context for the later case discussions in Chapters 3–5.

Policymaking in the EU

When it comes to combating climate change, the EU has long aimed for global leadership (Bretherton and Vogler, 2006; Gupta and Grubb, 2000; Oberthür and Roche Kelly, 2008; Schreurs and Tiberghien, 2007). Yet developments on the international stage have not always lived up to the notion of the EU leading and others following. This was made plain in the fifteenth Conference of the Parties (COP) to the United Nations Framework Convention on Climate Change (UNFCCC) in Copenhagen in December 2009 (Dubash, 2009), where EU negotiators were side-lined while final agreements were made behind closed doors among a small group of national leaders (Dimitrov, 2010; Falkner, Stephan and Vogler, 2010; Oberthür, 2011). Credible leadership requires action to be taken on the domestic level to demonstrate an ability and willingness to move forward, and to prove to the global community that, in the EU, action follows from rhetoric (Oberthür and Roche Kelly, 2008; Wurzel and Connelly, 2011a). Domestic/internal EU climate policy measures are thus necessary for credible international leadership, but also to preserve the EU's own self-interest, solidarity, and even survival (Behrens and Egenhofer, 2011: 219).

The process of agreeing on internal policy and legislation in the EU is complex, with many actors, institutions, preferences and interests to consider.

The EU has nevertheless managed to garner the political and institutional support to introduce several pieces of legislation aimed at combating climate change. Reaching agreement on a policy measure, and on the tools to achieve its goals, takes place among decision-making parties during the policymaking process – in the case of the EU's ordinary legislative procedure, among the European Commission ('the Commission'), the European Parliament ('the Parliament') and the Council of the European Union ('the Council').

I focus in particular on an analysis of CPI in the policy process and the policy output. Scholars of policy analysis have long tried to simplify the complex structure of government policymaking into a digestible and accessible form. Kingdon (2003: 2–3), in his influential study on agenda-setting in US policymaking, provides a definition of policymaking as a set of processes 'including at least (1) the setting of the agenda, (2) the specification of alternatives from which a choice is to be made, (3) an authoritative choice among those specified alternatives, and (4) the implementation of the decision'. In this book, the term 'policy process' in the EU context refers loosely to steps 1 and 2 of Kingdon's definition, while the 'policy output' refers to step 3. In other words, the policy process refers to the publication and negotiation on a policy proposal and the policy output refers to the final policy decision, or legislative act. The implementation of a policy measure (step 4) is not analysed here. Furthermore, in the context of this book, I do not discuss the nature or the purpose of the EU as a whole (but see, for example, Beck, 2009; Bretherton and Vogler, 2006; Damro, 2012; Manners, 2002). Instead, I view the EU as a supranational policymaking level that involves the three main decision-making institutions: the Commission, the Parliament and the Council. When analysing CPI in the policy process and output, I also acknowledge that these institutions do not operate within a vacuum, but interact with the politically important European Council and with many interest groups and lobbyists. Furthermore, policy development should be understood within a wider context of international events or developments that may also raise or lower policy priorities.

There are several different methods or procedures of policymaking in the EU, including: the ordinary legislative procedure, the intergovernmental method, and the open method of coordination. The intergovernmental method puts member states at the centre of decision making, with the Commission sharing the right of initiative, and the Parliament playing, at best, an advisory role. The open method of coordination does not in itself lead to binding legislation but aims to promote policy coordination and learning among member states (Lelieveldt and Princen, 2011), which can potentially lead to policy development in future. In this book, I focus on the ordinary legislative procedure, as, in each of the three cases examined later, policy measures were agreed under these rules. This procedure extends to more than 95 per cent of Community legislation (Lelieveldt and Princen, 2011: 67). Under this procedure, the Commission holds the sole right of initiative to make legislative proposals. Once the Commission publishes a policy proposal, it moves through negotiations between the Parliament and Council, with both these institutions enjoying full rights to amend and agree on the policy. Parliament adopts amendments by simple majority, with qualified majority voting (QMV) as the final decision-making rule in the Council (Art. 294 TFEU). Here, I discuss more specifically the role of each of the institutions involved directly in the EU's ordinary legislative procedure.

The Commission has a central role in the policymaking process, given its right as the initiator of policy proposals. The Commission is involved throughout the policy process, from initiation to agreement. It interacts with the Parliament and Council throughout the negotiations and oversees implementation. It plays a political as well as administrative role. As the EU has enlarged, the Commission has grown with it, and has developed its political skills in building consensus among 28 commissioners (Barnes, 2011: 43). Nevertheless, its ability to shape decisions in the EU is limited (and has been limited) by legal competence, its expertise and its size. In climate and energy policymaking, the Commission has moved from few, but radical, policy proposals in the early 1990s (such as the failed CO_2/energy tax) to new policy proposals in the late 1990s, to seizing opportunities for policy development that arose due to insufficiently ambitious past policies and external events. The Commission's Directorate-General (DG) for Environment was long responsible for climate policy, until DG Climate Action was created in 2009. Energy policy proposals are usually prepared by DG Energy. In 2014, Miguel Arias Cañete was named Commissioner for climate action and energy.

The Council brings together representatives of member states at the ministerial level. It operates in different sector-specific configurations. The transport, telecommunications and energy Council meets about six times a year and the environment Council meets about four times a year. The Council's role in policymaking includes acting as co-legislator with the Parliament. It can also, through the adoption of conclusions and recommendations, help push policy issues up or down the agenda, but it cannot move policy forward on its own (Lelieveldt and Princen, 2011; Oberthür and Dupont, 2011). Decision making in the Council under the ordinary legislative procedure follows the qualified majority voting rule (QMV). The number of votes per member state is weighted according to population, with no member state holding more than 29 (Germany, France, the UK and Italy) or less than three votes (Malta) (Corbett, Peterson and Bomberg, 2012: 58–59). QMV requires 55 per cent of the member states making up 65 per cent of the population of the EU, with a blocking minority required to represent more than 35 per cent of the population (TFEU, Art. 238.3).

The details of policy proposals are first negotiated at the level of working groups in the Council, before negotiations move to the Committee of Permanent Representatives (COREPER). The permanent representatives are the heads of the delegations of member states to the EU (McCormick, 2005: 90). In COREPER I, some of the more technical issues of a specific policy proposal are discussed among the member states' deputy permanent representatives. In COREPER II, the permanent representatives themselves meet to discuss the final political issues in the policy proposal, although not all policy proposals would be discussed in both COREPER I and II. With this elaborate system of preparation on a policy proposal, the Council meetings themselves do not discuss many policy proposals in-depth. A-point items on the Council meeting agenda are items that do not require further discussion after the preparatory meetings, and simply need the accord of the Council. B-point agenda items, however, call for further discussion and agreement at the ministerial level before they can be adopted. In reality, the majority of Council decisions fall under A-point agenda items, meaning the negotiations have taken place outside of the Council meeting itself (Lelieveldt and Princen, 2011: 61).

The Presidency of the Council rotates every six months among the member states, and provides member states with the opportunity to steer the policy agenda in line with their own priorities (Warntjen, 2007). Although the Presidency's room for manoeuvre may be limited, due to the inheritance of policy items already on the agenda and external events that require action, the Presidency can nevertheless play a facilitative role in pushing or blocking the adoption of policies (Lelieveldt and Princen, 2011: 60). In climate and energy policy, the Presidency could demonstrate political commitment to the climate issue by pushing forward on policies during its six-month mandate. This was the case with the Dutch Presidency in 1997, for example, under whose mandate the EU proposed to reduce GHG emissions by 15 per cent in time for the international climate negotiations in Kyoto, and with the French Presidency in 2008 that pushed for agreement on the climate and energy package under its mandate (Burns and Carter, 2011: 64; Oberthür and Dupont, 2011). In practice, the Council has historically often blocked or weakened the Commission's climate and energy policy proposals, especially in the 1990s and early 2000s (see below). As time went on, however, the Council began to demonstrate more political commitment to furthering climate policy and to energy policy development in the EU, motivated in part by ambitions to demonstrate leadership internationally.

The Parliament is the only elected institution in the EU, and thus most directly represents the European citizen in the policymaking process. It therefore is considered the most suited institution for voicing citizen concerns (Tanasescu, 2009: 48). It has gained formally equal power with the Council as a co-legislator in policies under the ordinary legislative procedure. It can also attempt to influence the policymaking agenda by adopting its own reports and resolutions, but it cannot push policy forward alone. Members of the European Parliament (MEPs) are organised along ideological (rather than national) lines in political groups. Much like in the Council, preparatory work on a policy proposal takes place outside the plenary of the Parliament. Parliamentary committees, composed of a number of MEPs from different political groups, prepare and debate the proposals before final decision-making takes place in the plenary sessions. The committee submits a report to the plenary that may (or usually does) contain proposed amendments to the Commission's policy proposal. In the plenary session, MEPs vote on each of the proposed amendments (Corbett, Jacobs and Shackleton, 2007; Corbett et al., 2012). The two committees most relevant for the research presented in this book are the industry, research and energy committee (ITRE) and the environment, public health and food safety committee (ENVI).

Traditionally, the Parliament is considered the most environmentally friendly of the three co-legislating institutions of the EU (Burns and Carter, 2011), although its ability to push for ambitious legislation is often limited by the continued political weight of the Council. In climate and energy policy development, the Parliament has often been overshadowed by the Council in negotiations, or Council's reactions have only weakly responded to Parliament's demands. For example, in the policy developments in the late 2000s on the climate and energy package, commentators remarked that Parliament's ability to push for ambitious policy was hampered by member state reticence. The role of Parliament in 2008 was rather to ensure that policy was adopted, than to push for more ambition on combating climate change. As Burns and Carter (2011: 69) highlighted: '[f]or the Parliament the passage of the climate change and energy package served to underline the limitations of its powers under co-decision when the Council is intransigent'.

The European Council is not one of the three decision-making institutions under the ordinary legislative procedure, but it is nonetheless an important political actor in the EU. With the entry into force of the Lisbon Treaty in 2009, the European Council became an official 'institution' of the EU (Art. 9). It brings together the heads of state and government (Prime Ministers or Presidents) of the member states at summit meetings organised about four times a year (twice per Presidency). Its importance as an institution stems from its political weight and its role and ability to provide the impetus for moving forward on specific issues. The 'Conclusions' coming from the meetings of the European Council often indicate the issues of political importance for member states at the time and can result in action from the Commission to initiate certain legislative proposals (Oberthür and Dupont, 2011; Tallberg, 2004; Warntjen, 2007).

In climate and energy policymaking, the conclusions of the European Council have come to provide the political impetus to develop more internal EU policy and to promote the EU's climate leadership ambitions internationally. In March 2007, the European Council agreed to an independent commitment to reduce GHG emissions in the EU by 20 per cent by 2020 compared to 1990 levels and called on the Commission to propose legislation (European Council, 2007). In October 2009, the European Council agreed to reduce GHG emissions by 80–95 per cent by 2050 (European Council, 2009). These political commitments provide a push for internal policymaking in the EU.

Beyond the three co-deciding institutions of the EU policymaking process, and the political impetus provided to the process by the European Council, external stakeholders and consulted institutions also play roles in the policymaking process. The European Economic and Social Committee (EESC) and the Committee of the Regions (COR) are usually consulted on policy proposals, but their opinions place no obligation on legislators. External stakeholders, lobbyists and interest groups attempt to influence policymaking, with lobbying efforts taking place at all stages of the EU policymaking process. The Commission and Parliament, especially, often seek out expert opinion on certain policy areas. These actors are accommodated in EU policymaking both formally (through official consultation procedures) and informally (through lobbying activities, personal relationships) (Hauser, 2011; Tanasescu, 2009; Watson and Shackleton, 2003).

But what are the main steps in the ordinary legislative procedure? The procedure is defined in Articles 289 and 294 of the TFEU. It begins with the Commission sending its policy proposal to the Parliament and Council for deliberation and negotiation. Both the Council and the Parliament have two opportunities (two 'readings') to discuss and agree on the proposal before the proposal is brought to a 'conciliation committee' at the third reading.

In the first reading, Parliament can adopt amendments (prepared by the committee in charge of the dossier, and voted on in plenary) to the Commission's proposal. The Commission provides an opinion on Parliament's amendments and can, if it wishes, put forward an amended proposal. Council's first reading can approve the Parliament's amendments, in which case the act is adopted, or not approve all the Parliament's amendments and/or suggest some of its own. In this latter case, the Commission provides an opinion on the Council's amendments and the proposal moves to the second reading.

In the second reading, there is a deadline to complete negotiations within three months (extendable by one month). Here the process follows a similar pattern as in the first reading. If Parliament agrees with the amendments of Council, the act is adopted. If Parliament rejects the Council's position, the act is not adopted. However, Parliament can also suggest new amendments to those of the Council, in which case the Commission provides a new opinion on these amendments and the proposal moves into second reading in the Council also. Again, if Council approves the new amendments put forward by the Parliament, the act is adopted. However, if the Council does not approve these new amendments, the proposal moves into the conciliation stage.

In the conciliation phase, the deadline for completing negotiations stands at six weeks, with the option to add an extra two weeks, after the conciliation committee is first convened. The conciliation committee's members come in equal number from the Council and the Parliament, with Commission representatives also attending the meetings. If negotiations in conciliation are unsuccessful, the act cannot be adopted. The committee sends its agreement to Council and Parliament for the third reading, and, if the proposal is adopted by both institutions (within the time deadline of six to eight weeks), it enters into law.

In addition, negotiations among the three institutions can occur in the form of 'trialogues'. These meetings take place in private and in an ad hoc fashion among representatives of the Parliament, Council and Commission. The aim is to reach a timely agreement on amendments that are acceptable to all deciding institutions. Although trialogues are generally first arranged during the second reading (European Parliament, 2012: 19), there are no fixed rules. A growing tendency to favour first reading agreements means that ad hoc informal negotiations are often ongoing from early stages. The trialogue meetings usually include the rapporteur from the Parliament in charge of preparing the lead committee's proposed amendments (and shadow rapporteurs from other political groups in the Parliament are occasionally present); the chairperson of COREPER I or the of the relevant Council working party (with assistance from the general secretariat of the Council); and representatives from the Commission, with the expert in charge of the dossier usually involved, along with representatives from the Commission's legal service. Recent studies on the evolution of decision making in the EU point to an increase in the informal negotiations among the institutions over time, and a corresponding increase in early or first reading agreements (De Clerck-Sachsse and Maciej Kaczynski, 2009).

EU climate and energy policy development

The European continent is not immune to the impacts of a changing climate. Human-induced climate change is caused by the emission of potent and long-lived GHGs into the atmosphere, the emissions of which have grown since pre-industrial times. As the most important GHG, carbon dioxide (CO_2) provides the benchmark against which all other GHGs are measured. The main sources of anthropogenic GHG emissions globally can be found in the burning of fossil fuels (oil, coal, natural gas) for electricity generation, transport, industry and in households. Deforestation, landfilling of waste, use of fluorinated gases and agricultural activities also represent significant, but smaller, proportions of GHG emissions. In the EU, about 80 per cent of GHG emissions are accounted for by energy production and consumption, including in transport, electricity, heating and cooling and consumption in industry and in households (EEA, 2010: 31). As European nations were among the first to undergo an industrial revolution in the late 1700s, they have historically contributed greater amounts of GHG emissions to the atmosphere, and the UNFCCC notes in the preamble that 'the largest share of historical and current global emissions of greenhouse gases has originated in developed countries'. Thus, when the UNFCCC was signed in 1992, the Convention stipulated 'common but differentiated responsibilities' among the signatories, and that developed country parties should 'take the lead in combating climate change and the adverse effects thereof' (Arts. 3.1 and 4.1). The picture has somewhat changed in the twenty years since then, with the EU's share in global GHG emissions declining as the share of rising emitters, such as China, soars.

As well as being an historical emitter of GHGs due to its early industrial revolution, Europe is expected to face differentiated impacts of the changing climate in the future. The main consequences of climate change that will be seen in Europe include 'an increased risk of coastal and river floods, droughts, loss of biodiversity, threats to human health, and damage to economic sectors such as energy, forestry, agriculture, and tourism' (EEA, 2010: 38). But these impacts are expected to be felt sooner and more severely in the Mediterranean, northwestern Europe, the Arctic and mountainous areas. Unless action is taken to mitigate climate change, greater costs will be involved in adapting to the impacts of climate change later (Stern, 2007). But EU action on climate change is also about political and international leadership, not just scientific and economic arguments. Despite these arguments, EU climate policy development was nevertheless slow to take off.

While the very foundation of the EU can be traced to links in energy policy across borders, climate policy has a much more recent history. The European Coal and Steel Community Treaty, signed in 1951, brought France, Germany, Italy, Belgium, Luxembourg and the Netherlands together in what would later expand and become the European Union. In 1957, the creation of Euratom, the European Atomic Energy Community, linked the European nations further, again with energy policy as a pillar of integration. Thus, the European integration project began on the strength of energy connections. Yet, since then, there has been limited competence for the European institutions to make policy on energy issues (Birchfield and Duffield, 2011). Climate change, on the other hand, emerged comparatively late in the development of the EU, but competence on the issue, based often on legal competences in the areas of the environment and the internal market, allowed the EU institutions to move forward in policy development.

In 1968, the Commission published a memorandum called 'first guidelines for a Community energy policy' (European Commission, 1968) in which it called for a common energy policy to counteract distortions in competition due to differences in energy costs from one member state to another. Already in 1968, the Commission highlighted the importance of an internal energy market, and the need to reduce member state dependence on imported energy. The oil crises in the 1970s (in 1973 and 1979) only heightened the risk of high-energy import dependence, yet even these shocks did not move member states towards more integration on energy policy. At this point, energy policy focused on the objectives of 'maintaining a regular, stable supply' of energy to the EU (European Commission, 1972: 8), and reducing energy import dependence. Energy efficiency and savings already emerged among the strategies for tackling security of supply issues. In 1974, the Council agreed to energy policy objectives to 1985 in response to the first oil crisis in the early 1970s. These targets included restricting dependence on imported energy to 50 per cent and stepping up energy savings. By 1978, it was already clear that further efforts would be required to meet these targets (European Commission, 1978) (see Table 1.1 for an overview of the main climate and energy measures from the 1950s to the 1980s).

Even in the face of the oil crises, little concrete progress was made in the 1970s to achieve an EU energy policy, despite Commission endeavours. The 1980s, however, saw a generally heightened environmental awareness, with linkages beginning to be made between energy and environmental policies. This followed both the 1972 United Nations Conference on the Human Environment in Stockholm and the World Climate Conference in Geneva in 1979 (Pallemaerts and Williams, 2006), which helped raise concern about acid rain, and (later) protecting the ozone layer and combating climate change. Attempts

Table 1.1 Some of the main policy developments in EU energy and climate policy, 1950s–1980s

Period	Energy policy	Climate policy	External events
1950s	**1951:** European Coal and Steel Community Treaty **1957:** European Atomic Energy Community	(Climate change regarded as a scientific issue)	
1960s	**1968:** First guidelines for a Community energy policy, COM(68) 1040	(Climate change regarded as a scientific issue)	
1970s	**1972:** Necessary progress in Community energy policy, COM(72) 1200 **1974:** – Community energy policy. Objectives for 1985, COM(74) 1960 – Towards a new energy policy strategy for the European Community, COM(74) 550 **1978:** Energy objectives for 1990 and programmes of member states, COM(78) 613 **1979:** The energy programme of the European Community, COM(79) 527	**1978:** Proposal for a multiannual research programme in the field of climatology, COM(78) 408	**1972:** United Nations Conference on Human Environment (Stockholm) **1973:** First oil crisis **1979:** – Second oil crisis – World Climate Conference, Geneva
1980s	**1981:** The development of an energy strategy for the Community, COM(81) 540 **1983:** – Proposal for a Council Regulation establishing specific measures of Community interest relating to energy strategy, COM(83) 31 – Community energy strategy: progress and guidelines for future action, COM(83) 305 **1985:** New community energy objectives, COM(85) 245 **1988:** The internal market for energy, COM(88) 238	**1986:** Parliament Resolution on the greenhouse effect **1988:** – The greenhouse effect and the Community, COM(88) 656 – Rhodes Council declaration on the environment	**1985:** Villach climate conference **1987:** Single European Act **1988:** IPCC established

to link climate and energy policy thus date from the emergence of climate change onto the political agenda in the late 1980s.

EU climate policy has rather developed alongside international agendas on climate negotiations since the 1980s. The Parliament was the first of the EU institutions to call explicitly for common policy measures to combat climate change (Jordan and Rayner, 2010). In 1986, the Parliament adopted its Resolution on 'measures to counteract the rising concentration of carbon dioxide in the atmosphere (the "greenhouse" effect)' (OJ C255/272 13.10.86). Furthermore, the adoption of the Single European Act (SEA) in 1987 included a specific requirement that environmental requirements 'shall be a component of the Community's other policies' (Art. 130r). In 1988, an international conference took place in Toronto on 'the changing atmosphere' (Pallemaerts and Williams, 2006: 23). The World Meteorological Organisation (WMO) and the United Nations Environment Programme (UNEP) agreed to set up the Intergovernmental Panel on Climate Change (IPCC) that same year, which aimed to engender consensus on climate science (Oberthür and Pallemaerts, 2010: 29). In the EU, the Commission released its first communication on climate change suggesting policy options to respond to the greenhouse effect (European Commission, 1988). The options included the recommendation that the EU and its member states take climate change into account in their policy decisions, especially related to energy (an early call for a certain degree of CPI); enhancing measures on energy savings and energy efficiency; and, sustaining research programmes on climate change (p. 11). The European Council took up the environmental concern in its Rhodes declaration on the environment (also in 1988). In this declaration, the European Council mentioned the greenhouse effect and provided political backing for action to combat climate change (European Council, 1988).

With the rise of the climate issue on the international and EU political agenda, and a general concern for environmental issues growing over the previous years, the Commission found the opportunity to develop further its competence on energy policy in making stronger links to environmental and internal market policy. Its 1990 Communication on 'energy and the environment' was the 'first time' that energy policy addressed 'environmental problems' in an international context (European Commission, 1990; interviw 22). In its 1995 green paper on energy policy (European Commission, 1995), the Commission outlined the 'trinity' of EU energy policy objectives (Pointvogl, 2009) as competitiveness, security of supply and environmental sustainability. These three objectives have remained at the core of EU energy policy since (see Table 1.2).

The Maastricht Treaty (1992) gave the EU competence on cross-border energy infrastructure, which was developed under the trans-European networks for energy (TEN-E) programme (Buchan, 2010: 360). But the legislative focus in energy policy in the 1990s was on developing the internal energy market, beyond infrastructure requirements – a process that would continue for many years. The first package of legislative measures came with the adoption of Directives

Table 1.2 Some of the main policy developments in EU energy and climate policy in the 1990s (with adopted Directives, Decisions and Regulations in italics)

Energy policy	Climate policy	External events
1990: Energy and the environment, COM(89) 369 **1993:** – *Council Directive 93/76/EEC to limit carbon dioxide emissions by improving energy efficiency (SAVE)* – *Council Decision 93/500/EEC concerning the promotion of renewable energy sources in the Community (ALTENER)* **1995:** An energy policy for the European Union, COM(95) 682 **1996:** – *Directive 96/92/EC concerning common rules for the internal market in electricity* – *Council Decision 96/391/EC laying down a series of measures aimed at creating a more favourable context for the development of trans-European networks in the energy sector* **1997:** – Energy for the future: renewable sources of energy, COM(97) 599 – The energy dimension of climate change, COM(97) 196 **1998:** *Directive 98/30/EC concerning common rules for the internal market in natural gas*	**1990:** – June: European Council called for adoption of climate targets – October: Joint Energy and Environment Council meeting called for stabilisation of CO_2 emissions in EU by 2000 **1992:** Proposal for package of measures to combat climate change including a proposal for a Council Directive introducing a tax on carbon dioxide emissions and energy, COM(92) 226 **1993:** *Council Decision 93/389/EEC for a monitoring mechanism of Community CO_2 and other greenhouse gas emissions* **1997:** Climate change – the EU approach for Kyoto, COM(97) 481 **1998:** – EU agrees on burden sharing to meet 8 per cent GHG emissions reduction target by 2008–2012 compared to 1990 levels – Towards an EU post-Kyoto strategy, COM(98) 353	**1990:** First IPCC report published **1992:** – UN Rio Earth Summit adopts UNFCCC – Maastricht Treaty **1996:** Second IPCC report published **1997:** – Kyoto Protocol (KP) adopted – Amsterdam Treaty

96/92/EC concerning common rules for the internal market in electricity and 98/30/EC concerning common rules for the internal market in natural gas, which both took several years to negotiate (Eikeland, 2011; Primova, 2015). Persistent problems of implementation, and the dominance of large incumbents in the energy market led to this legislation being amended (twice) in the 2000s. Besides these policy developments, some limited energy policy developments in the 1990s came in response to the challenge of climate change.

The EU continued to develop its climate policy alongside the international climate negotiations agenda. The early 1990s involved policy preparation leading to the 1992 Rio Earth Summit, where the UNFCCC was opened for signature. EU internal preparations for this conference involved (some) member states developing their own GHG emission reduction plans, but efforts were made at the EU level to develop common policies and measures also (Oberthür and Pallemaerts, 2010). In June 1990, the European Council called for the adoption of strategies and targets to limit GHG emissions, and for the Commission to develop proposals for concrete action to reduce emissions (European Council, 1990). In October 1990, the first joint Council of energy and environment ministers took place, and ministers responded to the European Council's call by adopting the target of stabilising CO_2 emissions by 2000 at 1990 levels (Council of the European Union, 1990). The Commission followed with its first package of proposals to combat climate change, including measures on energy efficiency (SAVE), on renewable energy (ALTENER), on monitoring CO_2 emissions, and on a CO_2/energy tax. This package was not adopted by the Council in time for the Rio conference in 1992 (Skjærseth, 1994: 31–32).

Of the four proposals, the CO_2/energy tax was by far the most controversial, with division even within the Commission on the issue, and opposition from several member states. Taxation Commissioner Christiane Scrivener was opposed to the measure and clashed with environment Commissioner Carlo Ripa di Meana on the issue (Skjærseth, 1994: 28). Although the proposal for a CO_2/energy tax survived discussions in the Commission, it then faced resistance in the Council, where unanimity was the required decision-making procedure. Several member states, and a strong business lobby, built opposition to the CO_2/energy tax. The proposal was blocked by Council and eventually withdrawn by the Commission in 2001 (European Commission, 2001). The EU's climate policy in the mid-1990s thus amounted to little more than a monitoring mechanism and two programmes on energy efficiency and renewable energies that had been substantially weakened in the policymaking process (Oberthür and Roche Kelly, 2008: 40).

In the run-up to international climate negotiations in Kyoto in 1997 there was again an internal push in the EU to demonstrate its leadership credentials on climate change by adopting targets and strategies, although this did not result in concrete policy developments. In June 1996, the environment Council declared the objective – in line with IPCC recommendations – that global temperature should not exceed 2° Celsius above pre-industrial level (Council of the European Union, 1996), an objective that has remained a guiding principle for climate policy since. The EU also proposed to reduce its emissions of CO_2, methane (CH_4), and nitrous oxide (N_2O) by 15 per cent by 2010 compared to 1990 in advance of the Kyoto negotiations, which was a far more ambitious proposal than other parties at the negotiations were prepared to suggest (Oberthür and Pallemaerts, 2010). In the end, the EU agreed to reduce its emissions of six GHGs – in addition to the three mentioned above, perfluorocarbons (PFCs), hydrofluorocarbons (HFCs) and sulphur hexaflouride (SF_6) – by 8 per cent. Meeting this goal meant negotiating a 'burden sharing agreement' among EU-member states to achieve the 8 per cent emissions reduction goal collectively. This agreement was reached in Council in June 1998 and adopted into law by Decision 2002/358/EC ratifying the Kyoto Protocol (Haug and Jordan, 2010; Lacasta, Oberthür, Santos and Barata, 2010). In all, the 1990s saw rather limited concrete movement forward on EU policies and measures in climate policy (see Table 1.2).

Achieving competence at the EU level on energy policymaking (at least for the internal dimension) seemed finally to be an attainable goal in the 2000s (see Table 1.3). The Commission produced another green paper on energy in 2000, this time titled 'towards a European strategy for the security of energy supply' (European Commission, 2000b). The Commission warned of the risks behind the growing energy import dependence of the EU and re-emphasised the three-pronged objectives of energy policy in the EU.

The EU adopted further legislation to take steps towards the completion of the internal energy market. The 'second package' was adopted in 2003 as Directives 2003/54/EC, repealing Directive 96/92/EC on electricity, and 2003/55/EC, repealing Directive 98/30/EC on natural gas, (Dupont and Primova, 2011; Eikeland, 2011). This step forward in the completion of the internal market opened up competition in electricity supply for all consumers by 2007. The 'third package' followed in 2009. Directives 2009/72/EC (on electricity) and 2009/73/EC (on natural gas) required the separation of supply and generation companies from transmission companies.

Along with the early energy and climate policy developments, including the 2001 Directive on the promotion of electricity produced from renewable energy sources (2001/77/EC; see Chapter 3) and the 2002 energy performance of buildings Directive (2002/31/EC; see Chapter 4), moves towards a common EU energy policy gained a boost after the informal European Council Hampton Court summit in 2005 (Buchan, 2009; Pointvogl, 2009; interview 22). This meeting was called by then UK Prime Minister Tony Blair during the UK Presidency, at a time when UK energy independence was diminishing and concern for climate change was increasing. Blair made energy policy a central element of the UK Presidency (Buchan, 2009: 8). Agreement was reached at this meeting to move forward on European energy policy and to establish a European grid. Soon after, in 2006, the EU suffered an energy security crisis when Russia and Ukraine engaged in a natural gas price dispute and gas supplies to Ukraine were cut. This dispute affected supplies to several EU countries – particularly the new member states in Central and Eastern Europe (Pirani, 2012).

Table 1.3 Some of the main policy developments in EU energy and climate policy in the 2000s (with adopted Directives, Decisions and Regulations in italics)

Energy policy	Climate policy	External events
2000: Towards a European strategy for energy supply security, COM(2000) 769 **2001:** *Directive 2001/77/EC on the promotion of electricity produced from renewable energy sources* **2002:** *Directive 2002/91/EC on the energy performance of buildings* **2003:** – *Directive 2003/54/EC concerning common rules for the internal market in electricity and repealing Directive 96/92/EC* – *Directive 2003/55/EC concerning common rules for the internal market in natural gas and repealing Directive 98/30/EC* **2004:** *Directive 2004/101/EC on the promotion of cogeneration* **2006:** – *Directive 2006/32/EC on energy end use efficiency and energy services* – A European strategy for sustainable, competitive and secure energy, COM(2006) 105 **2007:** An energy policy for Europe, COM(2007) 1 **2008:** Second strategic energy review: an EU energy security and solidarity action plan, COM(2008) 781 **2009:** *Directive 2009/28/EC on the promotion of the use of energy from renewable sources* **2010:** *Directive 2010/31/EU on the energy performance of buildings*	**2000:** Towards a European Climate Change Programme, COM(2000) 88 **2003:** *Directive 2003/87/EC establishing a scheme for greenhouse gas emission allowance trading within the Community* **2004:** – *Directive 2004/101/EC on the linking of the EU ETS with project mechanisms under the Kyoto Protocol* – *Decision No 280/2004/EC concerning a mechanism for monitoring Community greenhouse gas emissions and implementing the Kyoto Protocol* **2006:** *Directive 2006/40/EC on reducing the emission of fluorinated greenhouse gases* **2007:** European Council agrees 20 per cent GHG emissions reduction by 2020 target **2008:** – January: climate and energy package presented – December: European Council, Parliament and Council agree on climate and energy package **2009:** – *Directive 2009/29/EC amending Directive 2003/87/EC so as to improve and extend the greenhouse gas emission allowance trading scheme of the Community* – *Decision No 406/2009/EC on the effort of member states to reduce their greenhouse gas emissions to meet the Community's greenhouse gas emission reduction commitments up to 2020* – *Directive 2009/31/EC on the geological storage of carbon dioxide*	**2001:** – US withdraws from KP and EU agrees to pursue ratification – Third IPCC report published – Marrakech accords **2005:** – KP enters into force – Rejection of Constitutional Treaty in French and Dutch referenda **2006:** – Natural gas crisis (especially Eastern Europe) – Stern report published **2007:** – Fourth IPCC report published – Lisbon Treaty **2009:** Natural gas crisis (especially Eastern Europe) **2009:** COP15 in Copenhagen, Denmark **2010:** COP16 in Cancún, Mexico

Following the Hampton Court summit and the gas crisis in early 2006, the Commission released yet another green paper on energy policy discussing 'a European strategy for sustainable, competitive and secure energy' (European Commission, 2006). This green paper was followed closely by documents that outlined the future integrated climate and energy policies: the communications 'an energy policy for Europe' and 'limiting global climate change to 2 degrees Celsius: the way ahead for 2020 and beyond' (European Commission, 2007a, 2007b). EU climate policy development was now ever more closely linked to energy policy development.

In 2008, the Commission's second strategic energy review touched again upon the importance of energy security (European Commission, 2008), which showed a certain clairvoyance in the Commission in light of the ensuing gas crisis between Russia and Ukraine in early 2009 (see Chapter 5). However, the EU's competences on energy policy remain rather weak (Jordan and Rayner, 2010: 72). Article 194 of the TFEU opens a new energy chapter in the Treaty, but although competence is given to the EU on the energy market, security of supply, energy efficiency and renewable energy promotion and network interconnections, there remain two important caveats. These caveats are set out in paragraph two of the same Article: 'such measures shall not affect a member state's right to determine the conditions for exploiting its energy resources, its choice between different energy sources and the general structure of its energy supply', and in paragraph three: 'the Council ... shall unanimously ... establish the measures ... when they are primarily of a fiscal nature' (TFEU, Art. 194.2; 194.3). The procedural limits that helped block the CO_2/energy tax in the 1990s remain in place. The energy chapter seems to codify the policy developments that had already happened over the previous decades 'organically' through interlinkages with other competences (environment or internal market), rather than to specify potential new areas for EU energy policymaking (Buchan, 2009: 8–9).

The EU thus seems to have spurred action on energy policy in the 2000s based on previous developments. In the same decade, however, there was a great leap forward in climate policymaking. Climate policy internally in the EU in the 1990s was slow in being negotiated and often watered down substantially during the negotiations. But a renewed interest in internal climate policy development in the EU, especially in the mid-2000s, meant that often far-reaching policy could be proposed. This can in part be understood as the EU's attempts to make good on its claims for international leadership on climate change, but also due to a growing citizen concern for climate change (Boasson and Wettestad, 2013; Eurobarometer, 2008; Oberthür and Roche Kelly, 2008).

At the turn of the century, there was a sense of frustration in the Commission at the poor record and slow and difficult internal negotiations on climate (and energy) policy (Jordan and Rayner, 2010: 66). In 2000, the Commission tried a new approach with the proposal for a European Climate Change Programme (ECCP) that would involve many actors and stakeholders in developing measures and policies to combat climate change (European Commission, 2000a). During the first ECCP (from 2000 to 2004), eleven working groups, drawing on experts from the Commission, member states, industry and NGOs, aimed to identify options for reducing GHG emissions, covering topics such as emissions trading, energy supply and demand, transport, research and others.

On the international stage, a major upheaval in climate policy came in 2001 with the announcement from the George W. Bush administration that the US would not ratify the Kyoto Protocol. This led to much discussion in the EU about whether or not to move forward with ratification itself. Two criteria needed to be fulfilled before the Kyoto Protocol could enter into force: first, it needed to be ratified by at least 55 parties to the UNFCCC, and those parties needed to represent 55 per cent of industrialised countries' CO_2 emissions in 1990 (Art. 25.1, Kyoto Protocol). Without ratification by the United States, meeting these two criteria became more difficult. But the EU decided to work to save the Kyoto Protocol (Oberthür and Roche Kelly, 2008: 36). The EU was instrumental in ensuring the adoption of the Marrakech Accords on the implementation of the Protocol in 2001, and it ratified the Kyoto Protocol in 2002 (Damro, Hardie and MacKenzie, 2008: 180). Diplomatic efforts continued over the next number of years to ensure enough parties ratified the Protocol to ensure its entry into force. Once Russia ratified in late 2004 – after the EU promised to back its request for WTO membership – the Protocol finally entered into force in 2005 (Pallemaerts and Williams, 2006: 41).

During these early years of the twenty-first century, the centrepiece of adopted EU climate policy was the 2003 Directive on the development of the emissions trading system (ETS; Directive 2003/87/EC). This was a cap-and-trade scheme for CO_2 emissions from major industrial and energy installations in the EU. The cap (or ceiling) for emissions under the system was linked to the EU's commitment to reduce GHG emissions under the Kyoto Protocol (van Asselt, 2010: 128). The first (pilot) phase of the ETS ran from 2005–2007, and already demonstrated weaknesses in the system, due predominantly to an over-allocation of free emission allowances. Later reformulations of the ETS (see Table 1.3) aimed to reduce the cap and also the amount of free allowances provided to installations (Skjærseth and Wettestad, 2010; Wyns, 2015).

The EU entered a period of heightened political commitment to taking action on climate change from about the time of the entry into force of the Kyoto Protocol (Boasson and Wettestad, 2013: 43). A major political advancement for internal EU climate policy came with the European Council's endorsement of a goal to reduce GHG emissions in the EU by 20 per cent by 2020 in March 2007. While the European Council called on developed countries collectively to agree to reduce emissions 'in the order of 30% by 2020 compared to 1990', the EU made a 'firm independent commitment to achieve at least a 20% reduction of greenhouse gas emissions by 2020 compared to 1990'. In addition, the European Council agreed that the EU endorsed the 30 per cent target for itself by 2020 'provided that other developed countries commit themselves to comparable emission reductions and economically more advanced developing countries to contributing adequately according to their responsibilities and respective capabilities' (European Council, 2007: 12). This signalled a concrete political commitment from EU leaders to advance, not only internal EU climate policymaking, but also international climate negotiations.

The Commission's response to the commitment of the European Council was to propose a package of policy measures – the 'integrated' climate and energy package – on renewable energy, the ETS, carbon capture and storage (CCS) and effort sharing for sectors not covered by the ETS. Negotiations were rapid and agreement on the package was reached in the European Council in December 2008 (Boasson and Wettestad, 2013: 48; European Council, 2008). Council and Parliament also agreed to the measures in first reading in December 2008 and they were codified into EU law in 2009 (see Table 1.3).

The climate and energy package marked a period of high political commitment to action on climate change, and to promoting binding legislative measures (with the notable exception of the non-binding 20 per cent energy savings target to 2020). From 2005, with renewed emphasis on energy policy, the entry into force of the Kyoto Protocol, and the rejection of the EU's constitutional treaty in referenda in France (in May 2005) and in the Netherlands (in June 2005), a 'window of opportunity' to move forward on climate and energy policy was opened (Kingdon, 2003: 166). Climate change was seen to provide an opportunity to advance European integration generally, and to project EU leadership on the issue onto the international stage (Oberthür and Roche Kelly, 2008: 43).

Since the onset of the financial and economic crises in 2008, this window may have begun to close, and levels of political commitment to combating climate change do not seem as high as in the run-up to the 2009 international climate conference in Copenhagen. Climate and energy policy measures adopted since are few (the Energy Efficiency Directive, 2012/27/EU, being the most notable exception). As of 2015, however, the EU is preparing its climate and energy policy framework for 2030. In October 2014, the European Council adopted the target to reduce GHG emissions by 40 per cent by 2030, and this objective forms the basis of the EU's contribution to international climate negotiations due to take place in Paris at the end of 2015 (European Council, 2014). Policy proposals to bring the 2030 targets to light are in preparation, but are unlikely to be agreed as quickly as measures in 2008.

Structure of the book

This chapter presented the research set-up and set out the background and context for a study of CPI in the EU's energy policy. The main objective of the book is to explore the question: what is the extent of CPI in EU energy policy, and why? This exploration began here with a discussion of some essential background information on the policymaking process in the EU and on the historical development of EU energy and climate policy. In Chapter 2, I delve into the conceptual understandings of CPI and propose a methodology for measuring CPI in the policy process and output and an explanatory framework inspired by literature on EPI and on general theories of European integration. From there, I move to the analysis of the levels of CPI in the three case studies. Chapter 3 presents the renewable energy case. In Chapter 4, I discuss energy performance of buildings. And in Chapter 5, I present the external natural gas infrastructure case. Having measured the levels of CPI in each of the three cases, I then move, in Chapter 6, to apply the explanatory framework that is described in Chapter 2 to the results found. Finally, Chapter 7 brings out a number of broader conclusions from the research presented in the book. These include how the research contributes to ongoing scholarly debates, but also what the findings could mean for policymakers and stakeholders and the EU's ability to achieve its climate policy objectives. Considerable potential for future research on this topic exists, especially for research that could complement the findings and conclusions presented here and that could help refine the explanatory framework.

References

Adelle, C., Pallemaerts, M. and Chiavari, J. (2009). Climate Change and Energy Security in Europe: Policy Integration and its Limits. *SIEPS Report* (Vol. 4). Stockholm: Swedish Institute for European Policy Studies.

Adelle, C. and Russel, D. (2013). Climate Policy Integration: a Case of Déjà Vu? *Environmental Policy and Governance*, 23(1), 1–12.

Adelle, C., Russel, D. and Pallemaerts, M. (2012). A 'Coordinated' European Energy Policy? The Integration of EU Energy and Climate Change Policies. In F. Morata and I. Solorio Sandoval (eds), *European Energy Policy: an Environmental Approach* (pp. 25–47). Cheltenham: Edward Elgar.

Barnes, P. M. (2011). The Role of the Commission of the European Union: Creating External Coherence from Internal Diversity. In R. K. W. Wurzel and J. Connelly (eds), *The European Union as a Leader in International Climate Change Politics* (pp. 41–57). London: Routledge.

Beck, U. (2009). Understanding the Real Europe: a Cosmopolitan Vision. In C. Rumford (ed.), *The SAGE Handbook of European Studies* (pp. 602–619). London: Sage.

Behrens, A. and Egenhofer, C. (2011). Rethinking European Climate Change Policy. In V. L. Birchfield and J. S. Duffield (eds), *Towards a Common European Union Energy Policy: Problems, Progress, and Prospects* (pp. 217–234). New York: Palgrave Macmillan.

Bennett, A. (2008). Process Tracing: a Bayesian Perspective. In J. M. Box-Steffensmeier, H. E. Brady and D. Collier (eds), *The Oxford Handbook of Political Methodology* (pp. 702–721). Oxford: Oxford University Press.

Birchfield, V. L. and Duffield, J. S. (2011). *Towards a Common European Union Energy Policy: Problems, Progress, and Prospects*. New York: Palgrave Macmillan.

Boasson, E. L. and Dupont, C. (2015). Buildings: Good Intentions Unfulfilled. In C. Dupont and S. Oberthür (eds), *Decarbonization in the European Union: Internal Policies and External Strategies* (pp. 137–158). Houndmills: Palgrave Macmillan.

Boasson, E. L. and Wettestad, J. (2013). *EU Climate Policy: Industry, Policy Innovation and External Environment*. Farnham: Ashgate.

Bretherton, C. and Vogler, J. (2006). *The European Union as a Global Actor*. London: Routledge.

Briassoulis, H. (2005). Complex Environmental Problems and the Quest for Policy Integration. In H. Briassoulis (ed.), *Policy Integration for Complex Environmental Problems: the Example of Mediterranean Desertification* (pp. 1–49). Aldershot: Ashgate.

Buchan, D. (2009). *Energy and Climate Change: Europe at the Crossroads*. Oxford: Oxford University Press.

Buchan, D. (2010). Energy Policy: Sharp Challenges and Rising Ambitions. In H. Wallace, M. A. Pollack and A. R. Young (eds), *Policy-Making in the European Union*, 6th edn. (pp. 357–379). Oxford: Oxford University Press.

Burns, C. and Carter, N. (2011). The European Parliament and Climate Change. From Symbolism to Heroism and Back Again. In R. K. W. Wurzel and J. Connelly (eds), *The European Union as a Leader in International Climate Change Politics* (pp. 58–73). Abingdon, Oxon: Routledge.

Checkel, J. T. (2008). Process Tracing. In A. Klotz and D. Prakesh (eds), *Qualitative Methods in International Relations: a Pluralist Guide* (pp. 114–127). New York: Palgrave Macmillan.

Corbett, R., Jacobs, F. and Shackleton, M. (2007). *The European Parliament*. London: John Harper Publishing.

Corbett, R., Peterson, J. and Bomberg, E. (2012). The EU's Institutions. In E. Bomberg, J. Peterson and R. Corbett (eds), *The European Union: How Does it Work?*, 3rd edn. (pp. 47–73). Oxford: Oxford University Press.

Council of the European Union. (1990). *Conclusions*: Energy and Environment, October 1990.

Council of the European Union. (1996). 1939th Environment Council meeting, June 1996. *8518/96 Presse 188*.

Damro, C. (2012). Market Power Europe. *Journal of European Public Policy*, 19(5), 682–699.

Damro, C., Hardie, I. and MacKenzie, D. (2008). The EU and Climate Change Policy: Law, Politics and Prominence at Different Levels. *Journal of Contemporary European Research*, 4(3), 179–192.

De Clerck-Sachsse, J. and Maciej Kaczynski, P. (2009). The European Parliament – More Powerful, Less Legitimate? An Outlook for the Seventh Term. *CEPS Working Document, May 2009* (No. 314).

Dimitrov, R. S. (2010). Inside Copenhagen: the State of Climate Governance. *Global Environmental Politics*, 10(2), 18–24.

Dubash, N. K. (2009). Copenhagen: Climate of Mistrust. *Economic and Political Weekly*, XLIV(52), 8–11.

Dupont, C. and Primova, R. (2011). Combating Complexity: the Integration of EU Climate and Energy Policies. *European Integration Online Papers*, 15(Special mini-issue 1), Article 8.

EEA. (2010). *The European Environment: State and Outlook 2010 Synthesis*. Copenhagen: European Environment Agency.

EEA. (2014). *Trends and Projections in Europe 2014. Tracking Progress Towards Europe's Climate and Energy Targets for 2020*. Copenhagen: European Environment Agency.

Eikeland, P. O. (2011). EU Internal Energy Market Policy: Achievements and Hurdles. In V. L. Birchfield and J. S. Duffield (eds), *Towards a Common European Union Energy Policy: Problems, Progress, and Prospects* (pp. 13–40). New York: Palgrave Macmillan.

Eurobarometer. (2008). *Europeans' Attitudes Towards Climate Change*. Brussels.

Eurobarometer. (2011). *Climate Change. Special Eurobarometer 372*. Brussels.

Eurobarometer. (2014). *Climate Change. Special Eurobarometer 409*. Brussels.

European Commission. (1968). First Guidelines for a Community Energy Policy. *COM(68) 1040*.

European Commission. (1972). Necessary Progress in Community Energy Policy. *COM(72) 1200*.

European Commission. (1978). Energy Objectives for 1990 and Programmes of the Member States. COM(78) 613.
European Commission. (1988). Communication to the Council. The Greenhouse Effect and the Community. COM(88) 656.
European Commission. (1990). Communication from the Commission: Energy and the Environment. COM(89) 369.
European Commission. (1995). Green Paper: for a European Union Energy Policy. COM(94) 659.
European Commission. (2000a). Communication from the Commission on EU Policies and Measures to Reduce Greenhouse Gas Emissions: Towards a European Climate Change Programme (ECCP). COM(2000) 88.
European Commission. (2000b). Green Paper: Towards a European Strategy for the Security of Energy Supply. COM(2000) 769.
European Commission. (2001). Communication from the Commission: Withdrawal of Commission Proposals which are no longer Topical. COM(2001) 763/2.
European Commission. (2006). Green Paper: a European Strategy for Sustainable, Competitive and Secure Energy. COM(2006) 105.
European Commission. (2007a). Energy for a Changing World – an Energy Policy for Europe. COM(2007) 1.
European Commission. (2007b). Limiting Global Climate Change to 2 Degrees Celsius. The Way Ahead for 2020 and Beyond. COM(2007) 2.
European Commission. (2008). Second Strategic Energy Review: an EU Energy Security and Solidarity Action Plan. COM(2008) 781.
European Commission. (2011a). Communication from the Commission: Energy Roadmap 2050. COM(2011) 885/2.
European Commission. (2011b). Communication from the Commission. A Roadmap for Moving to a Competitive Low Carbon Economy in 2050. COM(2011) 112.
European Council. (1988). *Annex I: Declaration on the Environment*. SN 4443/1/88.
European Council. (1990). *Presidency Conclusions*, June 1990. Brussels: Council of the European Union.
European Council. (2007). *Presidency Conclusions*, March 2007. Brussels: Council of the European Union.
European Council. (2008). *Presidency Conclusions*, December 2008. Brussels: Council of the European Union.
European Council. (2009). *Presidency Conclusions*, October 2009. Brussels: Council of the European Union.
European Council. (2014). *Conclusions*, Document EUCO 169/14, October 2014. Brussels: European Council.
European Parliament. (2012). *Directorate-General for Internal Policies of the Union and Directorate for Legislative Coordination and Conciliations. Codecision and Conciliation: a Guide to How the Parliament Co-Legislates under the Treaty of Lisbon*. Brussels: European Parliament.
Falkner, R., Stephan, H. and Vogler, J. (2010). International Climate Policy after Copenhagen: Towards a 'Building Blocks' Approach. *Global Policy*, 1(3), 252–262.
George, A. L. and Bennett, A. (2004). *Case Studies and Theory Development in the Social Sciences*. London and Cambridge, MA: MIT Press.
Gerring, J. (2007). *Case Study Research: Principles and Practices*. Cambridge: Cambridge University Press.

Gerring, J. (2008). Case Selection for Case-Analysis: Qualitative and Quantitative Techniques. In J. M. Box-Steffensmeier, H. E. Brady and D. Collier (eds), *The Oxford Handbook of Political Methodology* (pp. 645–684). Oxford: Oxford University Press.

Gupta, J. and Grubb, M. (2000). *Climate Change and European Leadership*. Dordrecht: Kluwer Academic Publishers.

Hadfield, A. (2008). EU-Russia Energy Relations: Aggregation and Aggravation. *Journal of Contemporary European Studies*, 16(2), 231–248.

Haug, C. and Jordan, A. (2010). Burden Sharing: Distributing Burdens or Sharing Efforts? In A. Jordan, D. Huitema, H. van Asselt, T. Rayner and F. Berkhout (eds), *Climate Change Policy in the European Union: Confronting Dilemmas of Adaptation and Mitigation?* (pp. 83–102). Cambridge: Cambridge University Press.

Hauser, H. (2011). European Union Lobbying Post Lisbon: an Economic Analysis. *Berkeley Journal of International Law*, 29(2), 680–709.

Hermanns, H. (2004). Interviewing as an Activity. In U. Flick, E. von Kardoff and I. Steinke (eds), *A Companion to Qualitative Research* (pp. 209–213). London: Sage.

Hopf, C. (2004). Qualitative Interviews: an Overview. In U. Flick, E. von Kardoff and I. Steinke (eds), *A Companion to Qualitative Research* (pp. 203–208). London: Sage.

IEA. (2011). *World Energy Outlook 2011*. Paris: OECD/International Energy Agency.

IEA. (2013). *Tracking Clean Energy Progress 2013: IEA Input to the Clean Energy Ministerial*. Paris: OECD/International Energy Agency.

IPCC. (2007). *Climate Change 2007. Fourth Assessment Report: Synthesis Report*. Geneva: Intergovernmental Panel on Climate Change.

IPCC. (2013). Summary for Policymakers. In T. F. Stoker, D. Qin, G.-K. Plattner, M. Tignor, S. K. Allen, J. Boschung, … P. M. Midgley (eds), *Climate Change 2013: the Physical Science Basis. Contribution of Working Group I to the Fifth Assessment Report of the Intergovernmental Panel on Climate Change*. Cambridge: Cambridge University Press.

Jordan, A., Huitema, D., van Asselt, H., Rayner, T. and Berkhout, F. (2010). Governing Climate Change in the European Union: Understanding the Past and Preparing for the Future. In A. Jordan, D. Huitema, H. van Asselt, T. Rayner and F. Berkhout (eds), *Climate Change Policy in the European Union: Confronting the Dilemmas of Mitigation and Adaptation?* (pp. 253–275). Cambridge: Cambridge University Press.

Jordan, A. and Lenschow, A. (2010). Environmental Policy Integration: a State of the Art Review. *Environmental Policy and Governance*, 20(3), 147–158.

Jordan, A. and Rayner, T. (2010). The Evolution of Climate Policy in the European Union: an Historical Overview. In A. Jordan, D. Huitema, H. van Asselt, T. Rayner and F. Berkhout (eds), *Climate Change Policy in the European Union: Confronting the Dilemmas of Mitigation and Adaptation?* (pp. 52–80). Cambridge: Cambridge University Press.

King, G., Keohane, R. O. and Verba, S. (1994). *Designing Social Inquiry: Scientific Inference in Qualitative Research*. Princeton, NJ: Princeton University Press.

Kingdon, J. W. (2003). *Agendas, Alternatives, and Public Policies*, 2nd edn. London: Longman.

Klotz, A. (2008). Case Selection. In A. Klotz and D. Prakesh (eds), *Qualitative Methods in International Relations: a Pluralist Guide* (pp. 43–58). New York: Palgrave Macmillan.

Kvale, S. and Brinkman, S. (2009). *Interviews: Learning the Craft of Qualitative Research Interviewing*, 2nd edn. London: Sage.

Lacasta, N., Oberthür, S., Santos, E. and Barata, P. (2010). From Sharing the Burden to Sharing the Effort: Decision 406/2009/EC on Member State Emission Targets for

Non-ETS Sectors. In S. Oberthür and M. Pallemaerts (eds), *The New Climate Policies of the European Union: Internal Legislation and Climate Diplomacy* (pp. 93–116). Brussels: VUB Press.

Lafferty, W. M. and Hovden, E. (2003). Environmental Policy Integration: Towards an Analytical Framework. *Environmental Politics, 12*(5), 1–22.

Lelieveldt, H. and Princen, S. (2011). *The Politics of the European Union*. Cambridge: Cambridge University Press.

Lenschow, A. (ed.) (2002). *Environmental Policy Integration: Greening Sectoral Policies in Europe*. London: Earthscan.

Manners, I. (2002). Normative Power Europe: a Contradiction in Terms? *Journal of Common Market Studies, 40*(2), 235–258.

McCormick, J. (2005). *Understanding the European Union*. Houndmills: Palgrave Macmillan.

Nilsson, M. and Eckerberg, K. (eds) (2007). *Environmental Policy Integration in Practice: Shaping Institutions for Learning*. London: Earthscan.

Nollkamper, A. (2002). Three Conceptions of the Integration Principle in International Environmental Law. In A. Lenschow (ed.), *Environmental Policy Integration: Greening Sectoral Policies in Europe* (pp. 22–34). London: Earthscan.

Oberthür, S. (2011). The European Union's Performance in the International Climate Change Regime. *Journal of European Integration, 33*(6), 667–682.

Oberthür, S. and Dupont, C. (2011). The Council, the European Council and International Climate Policy: from Symbolic Leadership to Leadership by Example. In R. K. W. Wurzel and J. Connelly (eds), *The European Union as a Leader in International Climate Change Politics* (pp. 74–91). London and New York: Routledge.

Oberthür, S. and Pallemaerts, M. (2010). The EU's Internal and External Climate Policies: an Historical Overview. In S. Oberthür and M. Pallemaerts (eds), *The New Climate Policies of the European Union: Internal Legislation and Climate Diplomacy* (pp. 27–63). Brussels: VUB Press.

Oberthür, S. and Roche Kelly, C. (2008). EU Leadership in International Climate Policy: Achievements and Challenges. *International Spectator, 43*(3), 35–50.

Pallemaerts, M. and Williams, R. (2006). Climate Change: the International and European Policy Framework. In M. Peeters and K. Deketelaere (eds), *EU Climate Change Policy. The Challenge of New Regulatory Initiatives* (pp. 22–50). Cheltenham: Edward Elgar.

Persson, Å. (2004). *Environmental Policy Integration: An Introduction. PINTS – Policy Integration for Sustainability Background Paper*. Stockholm: Stockholm Environment Institute.

Pirani, S. (2012). Russo-Ukrainian Gas Wars and the Call on Transit Governance. In C. Kuzemko, A. V Belyi, A. Goldthau and M. F. Keating (eds), *Dynamics of Energy Governance in Europe and Russia* (pp. 169–186). Houndmills: Palgrave Macmillan.

Pointvogl, A. (2009). Perceptions, Realities, Concession – What is Driving the Integration of European Energy Policies? *Energy Policy, 37*, 5704–5716.

Primova, R. (2015). The EU Internal Energy Market and Decarbonization. In C. Dupont and S. Oberthür (eds), *Decarbonization in the European Union: Internal Policies and External Strategies* (pp. 25–45). Houndmills: Palgrave Macmillan.

Rathbun, B. C. (2008). Interviewing and Qualitative Field Methods: Pragmatism and Practicalities. In J. M. Box-Steffensmeier, H. E. Brady and D. Collier (eds), *The Oxford Handbook of Political Methodology* (pp. 685–701). Oxford: Oxford University Press.

Rohlfing, I. (2012). *Case Studies and Causal Inference*. London: Palgrave Macmillan.

Schreurs, M. A. and Tiberghien, Y. (2007). Multi-Level Reinforcement: Explaining European Union Leadership in Climate Change Mitigation. *Global Environmental Politics*, 7(4), 19–46.

Skjærseth, J. B. (1994). The Climate Policy of the EC: Too Hot to Handle? *Journal of Common Market Studies*, 32(1), 25–45.

Skjærseth, J. B. and Wettestad, J. (2010). The EU Emissions Trading System Revised (Directive 2009/29/EC). In S. Oberthür and M. Pallemaerts (eds), *The New Climate Policies of the European Union: Internal Legislation and Climate Diplomacy* (pp. 65–91). Brussels: VUB Press.

Stern, N. (2007). *The Economics of Climate Change: the Stern Review*. Cambridge: Cambridge University Press.

Swanborn, P. (2010). *Case Study Research: What, Why and How?* London: Sage.

Tallberg, J. (2004). The Power of the Presidency: Brokerage, Efficiency and Distribution in EU Negotiations. *Journal of Common Market Studies*, 42(5), 999–1022.

Tanasescu, I. (2009). *The European Commission and Interest Groups: Towards a Deliberative Interpretation of Stakeholder Involvement in EU Policy-Making*. Brussels: VUB Press.

Van Asselt, H. (2010). Emissions Trading: the Enthusiastic Adoption of an 'Alien' Instrument? In A. Jordan, D. Huitema, H. van Asselt, T. Rayner and F. Berkhout (eds), *Climate Change Policy in the European Union: Confronting the Dilemmas of Mitigation and Adaptation?* (pp. 125–144). Cambridge: Cambridge University Press.

Warntjen, A. (2007). Steering the Union. The Impact of the EU Presidency on Legislative Activity in the Council. *Journal of Common Market Studies*, 45(5), 1135–1157.

Watson, R. and Shackleton, M. (2003). Organized Interests and Lobbying in the EU. In E. Bomberg and A. Stubb (eds), *The European Union: How Does It Work?* (pp. 88–107). Oxford: Oxford University Press.

Wurzel, R. K. W. and Connelly, J. (2011a). Introduction: European Union Political Leadership in International Climate Change Politics. In R. K. W. Wurzel and J. Connelly (eds), *The European Union as a Leader in International Climate Change Politics* (pp. 3–20). London: Routledge.

Wurzel, R. K. W. and Connelly, J. (eds) (2011b). *The European Union as a Leader in International Climate Change Politics*. London: Routledge.

Wyns, T. (2015). Lessons from the EU's ETS for a New International Climate Agreement. *International Spectator*. doi:10.1080/03932729.2015.985941.

Yin, R. K. (2009). *Case Study Research: Design and Methods*, 4th edn. London: Sage.

2 From environmental to climate policy integration

In this chapter, I review the literature on policy coherence, coordination and integration and discuss the theories and explanatory variables used in environmental policy integration (EPI) literature (the main body of literature for a study on CPI). This leads to a conceptualisation of CPI and a suggested operationalisation of the concept for measurement purposes. Moving from operationalisation, I present a framework for explaining the levels of CPI that is found through empirical research. A study of CPI in the EU context can make use of general theories of European integration. I discuss briefly three theories from which I derive inspiration for a set of explanatory variables (as informed by the literature on EPI), including neofunctionalism, liberal intergovernmentalism and new institutionalism. The four main explanatory variables identified are: *functional interrelations; political commitment; institutional and policy context*, which each explain CPI in both the policy process and output; and *the process dimension*, which explains CPI in the policy output only.

To conceptualise CPI, it is necessary to situate the concept within the academic literature. Concepts such as policy coherence and policy coordination are linked to CPI. As climate policy has traditionally been part of the environmental policy field, CPI is particularly firmly rooted in the literature on EPI.

From policy coherence and policy coordination to policy integration

Policy coherence and policy coordination have long been promoted as tools for enhancing effective and efficient policymaking as part of wider policy analysis literature (Mickwitz et al., 2009; Peters, 1998; Sabatier, 2007; van Bommel and Kuindersma, 2008). Policy integration takes both these literatures into account and is a concept that engenders a holistic view of improving efficiency in policymaking (and effectiveness in the policy output and implementation). The literature around the concept of policy integration also has a relatively long history (Underdal, 1980, being a seminal early text on policy integration).

As key concepts within the policy integration discussion, 'policy coherence' and 'policy coordination' focus on and emphasise slightly different aspects of integration. Policy coherence aims to avoid conflicts between the objectives of different policy areas (and to promote win-win solutions as much as possible; van

Bommel and Kuindersma, 2008: 15). In other words, by promoting policy coherence, policy sectors would interact to agree on effective and efficient policies that produce successful combined results. Policy coherence thus emphasises the final policy output – aiming for policy outputs that are coherent with each other, and not contradictory or counterproductive (OECD, 2002).

In general, policy coordination can be seen to aim for the same results, namely that 'policies and programmes of government are characterised by minimal redundancy, incoherence and lacunae' (Peters, 1998: 296), although the emphasis is rather on achieving this through coordination efforts during the policy process. Peters has described policy coordination as an 'administrative Holy Grail' (1998: 295), implying that achieving policy coordination is a long-held challenge for policymakers. Metcalfe (1994: 281) suggests a scale of coordination that represents both a set of tasks for the achievement of policy coordination, and also a tool for identifying the level of policy coordination in any given polity. The scale ranges from low levels of policy coordination, where independent decision making in different policy sectors takes place, to high levels of policy coordination, where an overall government strategy is established. Policy coordination thus deals predominantly with improving the policy *process* and ensuring coordination takes place during the negotiation and policy development in and among policy sectors. The underlying assumption in policy coordination is that such efforts in the policy process will lead to coherent outputs, so the policy output flows from the policy process.

Policy coherence, coordination and integration are necessarily linked. Some analysts and researchers may contend that policy coordination and coherence are synonymous (Mickwitz et al., 2009: 24), but on further investigation, a clearer understanding is that policy coherence and coordination refer to different aspects of the policy cycle. Promoting policy coherence implies ensuring various policy outputs are harmonious; achieving policy coordination implies putting in place communication and coordination mechanisms to improve efficiency (leading to improved effectiveness) in the policy process. Achieving policy coherence in the policy output may be difficult without explicit policy coordination measures in the process. However, it may be possible to agree coherent policy outputs without specific coordination mechanisms, but due to overarching strategic directions. Similarly, policy coordination mechanisms do not necessarily guarantee a coherent output among several policies. Negotiating policy in a coordinated fashion across several policy 'ministries' or departments may raise some awareness about the various policy objectives, but this may still result in incoherent outputs. Research carried out on the policy process leading to the agreement of the Energy Efficiency Directive (Directive 2012/27/EU), for example, shows that coordination mechanisms within the European Commission did not prevent 'policy incoherence' in the policy output (Selianko and Lenschow, 2015: 20).

Policy integration, then, encompasses both coherence and coordination by emphasising the policy process *and* the policy output. Zingerli, Bisang and Zimmermann (2004: 3) also express the view that policy coordination is just one part of the overarching concept of policy integration – and can be regarded therefore as a 'degree of integration'. Policy integration is viewed as a combination of coherence and coordination to encompass the policy process and output (and also can be understood as a matter of degree itself, see below). The extent of policy integration overall may thus be established by examining both the policy process for evidence of the integration of the objectives of the second policy sector, and the policy output for synergy among policy objectives.

Arild Underdal developed a rationalist conceptualisation of policy integration, arguing that an integrated policy 'must meet three basic requirements, *viz* comprehensiveness, aggregation and consistency', corresponding respectively to the input stage (policy preparation), the processing of inputs (policy negotiations) and the output stage of the policy cycle (Underdal, 1980: 159). For the purposes of EU policy processes, we can understand the comprehensiveness in the input stage to mean the phase of the policy process where the European Commission assesses and proposes policy options (policy process). Aggregation in the processing of inputs refers to integration during the inter-institutional negotiations and amendments to the proposals of the Commission (policy process). Consistency in the output stage refers to the consistency (or coherence) between the multiple final policy decisions (policy output).

There is no clear agreement in the literature about the ideal end result of policy integration. For example, integration of two or more policies could eventually lead to the creation of one unified overall policy or, alternatively, certain policy concerns and objectives could be integrated into one or more existing policies. This lack of agreement is clear when Underdal is compared to Briassoulis. Underdal explains that an integrated policy means 'a policy where constituent elements are brought together and made subjects into a single, unifying conception' (1980: 159), promoting the idea that policies are integrated to become a new unified policy. However, Briassoulis disputes this single interpretation of the end goal of integration, highlighting in addition the aim of incorporating concerns into 'an extant policy to produce an integrated policy' (2005a: 50).

In related terms, policy integration may be conceived of as a unidirectional or mutual process of adaptation. Underlying these variances are frequently different conceptions of the weighting and prioritisation of separate policy objectives. Considering different policy objectives as equally valid may lead to pursuing mutual adaptation, the construction of an 'aggregate measure of utility' (Underdal, 1980: 162) and the creation of an 'integrated and coherent policy system' (Briassoulis, 2005a: 50). Giving priority to one particular policy objective may lead us to emphasise the unidirectional adaptation of integration into individual existing policies. These differences in approach also inform the debate about, and the conceptualisation of, environmental and climate policy integration.

From environmental to climate policy integration

Combating climate change has traditionally fallen under the remit of environmental policymaking, with climate policy development historically rooted in environmental concerns (although this focus in the climate debate has broadened considerably over the years). Broader environmental concerns have been linked to the pursuit of sustainable development, which promotes the integration of economic, social and environmental objectives for holistic sustainable policymaking (World Commission on Environment and Development, 1987). EPI literature stems from this pursuit of sustainable development, but highlights that special efforts are often required to ensure the environment is considered in policymaking (Lafferty and Hovden, 2003). The literature on EPI is well elaborated, and provides a conceptual setting for the further development of climate policy integration.

With the first development of EU environmental policy in the 1970s, the concept of environmental policy integration began to be discussed in EU policymaking. The Single European Act (SEA, 1987) introduced the objective that 'environmental protection requirements shall be a component of the Community's other policies' (Article 130r.2). EPI became an even more developed concept after the Amsterdam Treaty (1997) outlined sustainable development as an overall objective of the EU. In addition, Article 6 of the Amsterdam Treaty (now Article 11 of the TFEU) commits the EU by stating: 'environmental protection requirements must be integrated into the definition and implementation of the Union policies and activities, in particular with a view to promoting sustainable development'. This provides a legal basis for ensuring EPI, and it is thus a developed and familiar topic in the EU. However, while the concept of EPI has been enshrined as a legal objective of the EU, its practical application in policymaking has varied and often proven to be weak (Adelle and Russel, 2013; Bryner, 2012; Jordan and Lenschow, 2010).

The EU may well be the jurisdiction that has demonstrated most willingness and dedication to advance EPI, although its record has not resulted in strong standards of EPI in practice. The EU's first Environmental Action Programme (EAP), adopted in 1972, advanced environmental concerns such as the precautionary principle and the 'polluter pays' principle, leading to suggestions that commitment to responding to environmental concerns was present in the EU (European Commission, 1973; Lafferty and Hovden, 2003: 3). The Commission also discussed the notion of evaluating other policies for environmental effects in this EAP, and was the first EU institution to push for some form of EPI (Lafferty and Hovden, 2003: 3; Lenschow, 2002b: 3). However, even with this early commitment from the Commission, and the later codifications of the EPI principle into EU law in the Treaties, concrete responses to implementing EPI in policy were rather disappointing in the EU (Schout and Jordan, 2008: 162–164; Wilkinson, 1998, 2007). Early attempts to advance EPI included soft coordinating measures, yet throughout the 1980s and 1990s, very little advancement was made. The 1998 'Cardiff Process', which aimed to ensure that EU institutions (and especially the Council and the European Council) engaged in common reporting and policy reviewing for EPI, fizzled out without having lived up to expectations of advancing EPI (Jordan, Schout and Unfried, 2008: 163–165). Neither have institutions nor actors beyond DG Environment in the

Commission actively engaged with the objective of EPI in the EU. This may be partly explained by the conceptual vagueness of EPI, and the difficulties in defining and operationalising it (Lafferty and Knudsen, 2007). The literature on this topic is quite disparate – with different definitions, perspectives and explanatory variables. Indeed, EPI as a concept has thus far escaped definitional consensus and clarification.

EPI emerged and developed as a concept because of the perceived failure and inability of traditional sectoral environmental policymakers to respond to the pressures placed on the environment by the actions of a wider society. It can be considered a broad concept that is value-laden and political, thus resulting in several interpretations (Persson, 2007: 44). There is thus much debate on the meaning of the term, how it can be defined, recognised, implemented, measured, evaluated and improved upon. In fact, the European Environment Agency describes how the lack of agreement on the meaning of EPI has helped to make it 'more acceptable to policy-makers' (EEA, 2005: 12). What is clear, however, is that EPI has a normative dimension in favour of the environment. Placing an adjective before the term 'policy integration' implies assigning preference, weight or priority to a specific policy sector's objectives over another (Briassoulis, 2005b: 23). As outlined above, it is clear that European law reflects this normative dimension, supporting in particular the integration of environmental objectives into other policy sectors.

What remains unclear is *how much* weight or priority the environment does or should receive in the policy process and its output. Reflecting the broader discussions about policy integration in general, authors differ on whether they regard EPI as a policy principle; a process or a set of measures or institutional procedures; a policy output or outcome; or some combination of these (Persson, 2004: 22–25). In this context, the variety of strength and depth of EPI employed in the literature alters the measurement of EPI against a particular benchmark. Some have advocated that environmental objectives should receive 'principled priority' in other policy sectors (Lafferty and Hovden, 2003: 9), throughout the policy process and, especially, in the final output (also called 'strong' EPI). On the other end of the spectrum, others have emphasised the importance of simply taking environmental considerations into account in the formulation of policy in other sectors, which can be considered 'weak' EPI (Jordan and Lenschow, 2008b; Persson, 2004).

In conceptualising CPI, I take inspiration from the 'strong' standard set by Lafferty and Hovden in the policy process and output. Their two-tiered definition of EPI is as follows:

- the incorporation of environmental objectives into all stages of policymaking in non-environmental policy sectors, with a specific recognition of this goal as a guiding principle for the planning and execution of policy;
- accompanied by an attempt to aggregate presumed environmental consequences into an overall evaluation of policy, and a commitment to minimise contradictions between environmental and sectoral policies by giving principled priority to the former over the latter (Lafferty and Hovden, 2003: 9).

Bryner provides a reasoned (and vague) definition of EPI as 'a greater awareness of how policies in one area affect another [that] encourages policymakers to prioritise environmental concerns' (2012: 22). Even this looser definition provides a normative understanding in favour of the environment. Nevertheless, there are good reasons for accepting the unavoidably normative standard for EPI (and CPI) of 'principled priority' as a benchmark. Taking a stance is inescapable due to the inherently normative nature of the concepts of EPI and CPI as standards of policy evaluation. Most importantly, the standard for operationalising CPI needs to be explicated to make it transparent, to facilitate comparison with other research results, and to enable criticism and a full understanding. Choosing a high standard (such as 'principled priority') for CPI has the advantage of being comprehensive in two dimensions. First, it covers the policy process and its output, and second, it can arguably cover the full spectrum from strong to weak EPI. As we will see below, in operationalising CPI, a principled priority standard provides a benchmark for evaluation, which can lead to an understanding of the gap between the status quo and the benchmark of strong CPI.

On the basis of the above discussion of EPI, I take the strong definition of CPI as a benchmark for measurement – therefore CPI can indeed be considered a matter of degree (Bryner, 2012). CPI can exist to lower or higher levels in the policy process and policy output of sector policies and an understanding of the concept of CPI that allows for the principled priority of climate policy objectives over the other policy sector objectives is rather an expression of a high benchmark against which actual levels can be measured. Therefore, rather than defining CPI as assigning principled priority in policymaking, I suggest that the principled priority of climate policy objectives in all stages of the policy process and its output in non-environmental policy sectors is the highest level of CPI that can be achieved.

Additionally, principled priority as a definitional concept for CPI is problematic for sustainable development objectives. While taking Lafferty and Hovden's definition of EPI as outlined above, and simply replacing the word 'environment' with 'climate' results in a reasonable definition (Ahmed, 2009: 11), this is problematic and counterproductive if the aim is to prevent irreversible damage to environmental life-support systems. For example, a policy that responds to the climate challenge by promoting fast-growing forests may detrimentally affect biodiversity and habitat protection. Consequently, some literature has tried to overcome the problem of assigning principled priority to one environmental objective over another by distinguishing between what has come to be called 'external' EPI in non-environmental policy sectors, and 'internal' EPI between environmental sub-policies. In this way, principled priority can be granted to environmental/climate objectives for cases of external EPI, while a balancing and synergy logic is applied to cases of internal EPI (Biermann, Davies and van der Grip, 2009; Kulovesi, Morgera and Muñoz, 2010; Oberthür, 2009). As I focus in this book on 'external' CPI into the EU's energy sector, the strong standard of 'principled priority' can appropriately be applied as a benchmark of policy assessment.

Operationalising CPI

The empirical application of the strong CPI standard requires an understanding of how to measure different levels of CPI, in both the policy process and its output. For the purposes of such a measurement, an overall guiding question can be asked in each case: what would the very high, principled priority level of CPI in this sector look like? This question is examined for both the policy process and the policy output. I suggest a fivefold scale to measure (qualitatively and quantitatively) the level of CPI, from very low to very high over low, medium and high (see Tables 2.1 and 2.2). In measuring CPI in the policy process, I use a number of factors and sub-indicators to arrive at the final result, while for CPI in the policy output I examine how the policy output under examination closes the gap between business-as-usual scenarios and required climate policy objectives to achieve decarbonisation by 2050.

Under the ordinary legislative procedure, it is possible that very high levels of CPI would be found where there is ease of access of pro-climate policy stakeholders to the policy process. The involvement of pro-climate policy stakeholders in the policy process is an indicator of the attention paid to the climate policy issue during policymaking. In much EPI literature, emphasis is laid on the openness of the policy process to several voices to ensure environmental/ climate issues are heard and dealt with (EEA, 2005; Jordan and Lenschow, 2008a; OECD, 2002). Hence, the involvement of pro-climate stakeholders is considered here as one main indicator of the level of CPI in the policy process, which is broken down into three distinct operational indicators.

First, I assess the role played by internal climate policy stakeholders in the policy negotiations. This refers to the roles played by DG Environment (and later, DG Climate Action, if applicable) in the Commission, by the Environment (ENVI) committee in the Parliament, and by the environment Council throughout the policy negotiations. To assess this, I look at the responsible DGs, committees and Council formations on each policy file, and if these are not the pro-climate policymakers, then assess whether or not these pro-climate actors were involved in the negotiations (and to what extent). These roles vary loosely on a qualitative fivefold scale, matching the scale for the level of CPI in the policy output, from 'very high' (when internal pro-climate stakeholders are in the lead on the policy proposal and negotiations) to 'very low' (when there is practically no involvement and no visibility of pro-climate stakeholders in the process). The fivefold scale is thus based on the conceivable possibilities for pro-climate stakeholder involvement in a certain policy file within the institutions. The different levels of 'involvement', in this case, are dictated by the procedural rules that could be applied in the EU's ordinary legislative procedure. Through an analysis of documents associated with the policy negotiations, supplemented with interview data, media reports and literature review, I can assign a qualitative score on the 'involvement of internal pro-climate policy stakeholders' within the EU co-deciding institutions in the policy process (see Table 2.1).

Second, I look to the role played by external pro-climate policy stakeholders in the policy process – such as environmental NGOs and industrial actors in favour of stringent climate policy. The involvement of external pro-climate actors in the policy process can lead to heightened pressure on policymakers to ensure climate policy objectives are integrated into the development of policy (EEA, 2005; Lenschow, 2002a; Persson, 2007). Access to the deciding institutions can depend on two factors – first, the procedures for consultation that are in place in a certain institution, and second, the 'openness' of the institution to input from external stakeholders. These two factors provide the basis for the five-fold scale for the measurement of the involvement of external pro-climate stakeholders in the policy process. This is assessed through an analysis of the official public consultations organised by the EU institutions, but also backed-up qualitatively by interviews with involved stakeholders who assess their own involvement in the policymaking process, with media reports and with a literature review. Thus, a qualitative score of involvement of external stakeholders can be assigned from 'very high' (pro-climate stakeholders are invited by the EU institutions to provide input in the policy process; access is guaranteed and easy; and pro-climate stakeholders' arguments receive wide backing) to 'very low' (no pro-climate stakeholders involved or visible in the policymaking process; see Table 2.1).

The two above indicators are based on a number of assumptions. First, linking access of stakeholders to influence is particularly difficult methodologically (Hauser, 2011). However, as a measurement of CPI in the policy process, having some access to the process should at least allow for climate arguments to be raised. The strength of climate voices may also depend on the messages coming from other stakeholders that are not in favour of CPI. Thus, carrying out an analysis of all stakeholders involved in each case may provide richer data. In the EU's energy policy sector, recent studies argue that stakeholders come predominantly from large, traditional energy companies that may not support CPI (Vasileiadou and Tuinstra, 2013). Such an in-depth analysis of stakeholders is unfortunately beyond the scope of this book. By focusing on multiple methods of qualitative data collection, including interviews with pro-climate stakeholders who assess their own involvement, it is hoped that the levels of CPI as measured in the processes of the cases can nevertheless be considered roughly accurate. A second assumption is that stakeholders that are identified as pro-climate stakeholders may not be the only actors convinced of the importance of acting on climate change. Indeed, policymakers in the energy sector may already be convinced of the need to push CPI. If none of the identified pro-climate stakeholders is involved in the process, the result of CPI in the process would then be artificially low. However, the third indicator (see below) – the recognition of climate policy objectives by policymakers – should alleviate some of the problems of identifying the extent to which even non-climate stakeholders are acting in favour of CPI.

The third indicator for the measurement of the extent of CPI in the policy process includes the treatment of functional interactions with climate policy objectives in the policy discourse. Where actors in a policy sector recognise the potential interrelations between long-term (to 2050) climate policy objectives and their policy developments, and where these interrelations are continually recognised and referred to throughout the process, this can help lead to higher levels of CPI in the policy process (and, potentially also in the policy output). Measuring this indicator includes a document analysis of policy papers, literature and media reports throughout the process to assess whether scientific arguments (from officially recognised bodies such as the IPCC) about the need to combat climate change are referred to throughout the development of a policy measure (on the importance of scientific input for policy integration, see, for example Jacob, Volkery and Lenschow, 2008; OECD, 2002). This analysis is supplemented with interviewee opinions on the relative importance of climate change as a motivating factor in the policy development (demonstrating actor understanding of functional interrelations in the policy discussions). In this case, CPI is considered 'very high' if policymakers recognise and articulate the functional interrelations between their sector and long-term climate policy, and aim to combat climate change through scientifically-grounded policy measures. At the lowest end of the fivefold scale, 'very low' CPI would be found if the interrelations with climate policy objectives are not recognised or are ignored in policy development. This fivefold scale ensures a sufficiently differentiated understanding of the level of recognition of functional interrelations by policymakers in the EU, without imposing too-rigid or too-flexible a scale (see Table 2.1).

Taking an aggregate measurement of the three indicators leads to an overall single score for the extent of CPI in the policy process. Table 2.1 summarises the operationalisation of CPI in the policy process. The aggregate score does not distinguish between the three indicators in terms of importance – they are equally weighted. This rough estimation may become more refined in future as new applications of such an operationalisation of CPI in different policy sectors take place.

In terms of policy output, very high levels of CPI will be achieved if policies are close to 100 per cent in line with established (and scientifically-grounded) climate policy objectives. This means that policies must aim to achieve the long-term climate objective of reducing GHG emissions by between 80 and 95 per cent by 2050 in the EU. This particular goal is supported by the scientific reports of the IPCC (IPCC, 2007, 2013) and also received political backing in the European Council meeting in October 2009 (European Council, 2009). It is a translation from long-standing global objectives to limit temperature increase to 2° Celsius. Therefore, not only is this objective scientifically grounded, but it is also a stated objective of the EU as a whole.

With this standard in mind, an investigation of the levels of CPI in the policy output can establish the extent of the gap between the policy output of a particular policy measure and the very high level of CPI. For example, in renewable energy (RE) policy, a goal of achieving between 80 and 100 per cent share of renewables in EU energy consumption by 2050 can be considered as in line with 2050 climate policy objectives (see Chapter 3; EREC, 2010; European

Table 2.1 Measuring the level of CPI in the policy process

Level: Indicator:	Very low CPI	Low	Medium	High	Very high CPI
1. Internal pro-climate stakeholders	No involvement and no visibility in the policy process	Not consulted, but providing unsolicited opinions	Consulted throughout the process	Co-drafters and co-deciders with other sectoral DGs, committees and Council formations	Leading on the policy proposal development and negotiations
2. External pro-climate stakeholders	No pro-climate stakeholders involved or visible in the policymaking process	No or little access to the policy process, pro-climate stakeholders' arguments receive general opposition	Procedures allow access, but ease of access limited, pro-climate lobby facing certain opposition from other stakeholders and/or policymakers	Easy access to policymaking, strong pro-climate lobby with little opposition from other stakeholders and/or policymakers	Invited and opinions sought by policymakers, access guaranteed, pro-climate stakeholders' arguments receive general support
3. Recognition of interrelations with climate policy objectives	Functional interrelations with climate policy objectives not recognised	Climate policy objectives mentioned, yet interrelations not strongly featuring throughout policy process	Functional interrelations recognised and considered throughout policy process, yet without motivating policy action	Achieving climate policy objectives is a stated goal of the policy and continues to motivate policy development throughout the process	Policy is developed in order to achieve long-term climate policy objectives and such a standard remains in place throughout the policy process

Commission, 2011). This policy goal is considered evidence of very high CPI (see Table 2.2). However, if the RE policy output under examination achieves less than an 80 to 100 per cent share of RE in the EU by 2050, I measure the distance between this very high standard and the actual output. Nonetheless, as most policies do not put in place specific measures to 2050, it is rather by assessing the *pathway* that the policy output sets towards the 2050 goals that I measure the level of CPI in the policy output. Taking again the example of RE policy, the EU's 2020 goal of increasing the share of RE in the EU to 20 per cent is a policy

Table 2.2 Measuring the level of CPI in the policy output

CPI in the policy output:	Very low CPI	Low	Medium	High	Very high CPI
	0–20%	21–40%	41–60%	61–80%	81–100%

output that may or may not put the EU on the pathway to achieving the 2050 climate goals. By examining the pathway that very high levels of CPI in the policy output would require, I identify the sort of policy output for 2020 that would lead us to this level. The distance between this and the actual policy output to 2020 (measured as a percentage of the policy gap) represents the level of CPI found in EU RE policy to 2020. Thus, on this basis, I assess the level of CPI in the policy output by again applying a fivefold scale ranging from very low over low, medium and high to very high (see Table 2.2).

For the purposes of establishing the contribution of the various energy policies to climate policy objectives in the later chapters, I refer specifically to a collection of roadmaps and scenarios to 2050 as well as to literature on decarbonisation (Dupont and Oberthür, 2015b). These include the European Commission's 2050 energy roadmap and its impact assessments (European Commission, 2011); the European Climate Foundation's selection of scenarios to 2050 (ECF, 2010); the European Renewable Energy Council's (EREC's) 'RE-thinking 2050', and Greenpeace and EREC's 'energy [r]evolution' (EREC and Greenpeace, 2010; EREC, 2010); Eurelectric's 'power choices' (Eurelectric, 2010); Eurogas' roadmap to 2050 (Eurogas, 2011); WWF's 'energy report' (WWF, 2011); Friends of the Earth's report on 'Europe's share of the climate challenge' (Heaps, Erickson, Kartha and Kemp-Benedict, 2009); and the IEA's 2011 world energy outlook, among others (IEA, 2011a, 2011b). Not all of these roadmaps and scenarios are directly comparable due to different sectors of focus, different objectives and methodologies. However, taking them together, a general picture of the possible (and desirable) goals for climate policy for 2050 can be drawn for each of the cases discussed in Chapters 3–5 (Dupont and Oberthür, 2015a; Lechtenböhmer and Samadi, 2015). In each case, reference is made to these studies and scenarios in order to establish the benchmark of very high levels of CPI against which the policy reality is measured.

Explaining climate policy integration

Having established what CPI means and how it can be measured in the policy process and output, it is time to turn to the explanatory framework: what variables can be identified that will help us explain and understand the levels of CPI found?

To develop such an explanatory framework, I review the explanatory variables set out in general EPI literature and in several theories of European integration that, combined, can help provide a greater understanding of the EU policy-making process generally and the extent of CPI specifically (Rosamond, 2007;

Warleigh-Lack and Drachenberg, 2010; Wiener and Diez, 2009a). I draw inspiration from broader theories of European integration to bring various explanatory variables highlighted in the EPI literature into a manageable framework. I do not aim to carry out a theory-testing exercise but a study of the empirical reality of CPI in EU policymaking. Thus, the combination of several theoretical perspectives, as promoted notably by Wiener and Diez in 2009, can help ensure as comprehensive an understanding of empirical results as possible (such as combining state- and process-centred perspectives, and not rejecting the role played by institutions and actors in EU policymaking). Such a conceptual combination of perspectives is increasingly promoted also in international relations theory as a pragmatic way of advancing knowledge (see, for example, Friedrichs and Kratochwil, 2009).

EPI literature provides a disparate and long list of potential explanatory variables, depending largely on the conceptualisation of EPI. Whether EPI is viewed as an overarching legal perspective, a tool for efficient policymaking, a policy outcome or an aspirational goal, differing explanatory variables come to the fore. Few examples of truly comprehensive explanatory frameworks have evolved (Adelle and Russel, 2013; Dupont and Oberthür, 2012; Jordan and Lenschow, 2010; Persson, 2004). Explanatory variables in EPI literature also depend largely on the focus of analysis. Whether the focus of the analysis is on the policy process or output, or on the overarching principle of EPI, different explanatory variables have been emphasised. For example, research focusing on the policy process has employed an institutional perspective, emphasising coordination mechanisms (Jordan and Lenschow, 2008b), or a 'policy learning' perspective highlighting the importance of policy framing (Nilsson, Eckerberg, Hagberg, Swartling and Söderberg, 2007; Nilsson and Persson, 2003). A legal perspective has been employed to assess the overarching legal commitment to EPI (Nollkamper, 2002), while policy evaluation studies have attempted to assess EPI in the policy outcomes, emphasising resources, monitoring of policies and political commitment (EEA, 2005; Lafferty and Knudsen, 2007; OECD, 2002). Different theories have thus been used to understand the varied rates of EPI in different jurisdictions, including governance theories (von Homeyer, 2006), learning theories (Nilsson and Persson, 2003) and theories of bureaucratic politics (Lenschow, 2010). Each of these perspectives helps to explain part of the EPI story, although most EPI research in the past has focused specifically on the policy process (Adelle and Russel, 2013; Dupont, 2011).

As this is a study on the extent of CPI in the EU, the explanatory variables highlighted in the EPI literature can be linked to the insights on EU policymaking found in general theories of European integration. As a result of the review of EPI literature and theories of European integration, four main explanatory variables are defined and discussed below: *functional interrelations; political commitment; institutional and policy context*, which each explain CPI in both the policy process and output; and the *process dimension*, which explains CPI in the policy output only (Dupont and Oberthür, 2012; Dupont and Primova, 2011).

Drawing inspiration from European integration theories

General theories of European integration have been in development for as long as the European project has existed. The number of theories has grown, while their scope has often narrowed. The so-called 'grand theories' of European integration include neofunctionalism and liberal intergovernmentalism (earlier, neorealism). Other theoretical perspectives do not aim towards explaining or understanding the European integration project as a whole, but rather certain aspects of the European polity, policymaking processes or the politics of the EU (Wiener and Diez, 2009b: 245). In addition, new formulations of the 'grand theory' of neofunctionalism have reduced its scope to a theory with only partial explanatory power (Stone Sweet and Sandholtz, 1998). For understanding EU governance and policymaking, the various theoretical perspectives provide insights that emphasise different actors and processes.

Rather than seeing the theories as representing conflicting views and perspectives, I follow the understanding of Diez and Wiener that the theoretical approaches can be seen as 'adding to a larger picture, without being combined into a single, grand theory' (2009: 17). In other words, in order to gain an understanding and explanation of (CPI in) EU policy processes and outputs, a combination of several theoretical perspectives is useful to aim for a comprehensive understanding. Therefore, I draw inspiration from a combination of three theories to develop the explanatory framework. These are two so-called 'grand' theories of European integration (neofunctionalism and liberal intergovernmentalism), and one more middle-ground theory (new institutionalism). These theories are particularly helpful for establishing a fuller explanation of the extent of CPI, due to their different areas of focus in the EU policymaking process.

Neofunctionalism provides explanations with reference to the involvement of multiple actors in the policymaking process, information exchange and the functional interlinkages between policy sectors. Liberal intergovernmentalism's focus on political bargains and member states as the most important actors helps explain CPI with reference to the level of political commitment and leadership to climate change and CPI generally, and on the role of the Council and the European Council. Finally, institutionalism can explain CPI by highlighting the importance of institutions, the policy and institutional context of the decision-making processes, and the legacy of previous policies and decisions. Although no one perspective seems suited to explain the level of CPI comprehensively, connections among the theoretical perspectives and the explanations they provide can be expected. Thus, since each theory focuses on these different aspects of the policymaking process and with different assumptions about the drivers of European integration more generally, there is a case to be made in deploying these perspectives in a compatible way, without ascribing a hierarchy of importance to the various perspectives. In this way, the theories become tools for understanding the empirical results in a particular EU policy domain, as opposed to ideological frames for describing the overarching European integration project.

Neofunctionalism dates from the 1950s and 1960s, when Ernst Haas and Leon Lindberg elaborated the theory to explain the evolution of the European Coal and Steel Community (Haas, 1958; Lindberg, 1963). The roots of this theoretical perspective can be found in functionalist and federalist thought, which emphasise integration through day-to-day decision making, incremental changes and learning processes (Niemann and Schmitter, 2009). In its early development, neofunctionalism aimed towards grand theorising of European integration, but by focusing on the nitty-gritty of day-to-day decision making that leads to 'spillover' in other policy domains and increased integration (Diez and Wiener, 2009; Haas, 1958; Niemann and Schmitter, 2009; Strøby-Jensen, 2007). In the 1970s and later, however, the theory underwent reformulation and revision, with neofunctionalist scholars disagreeing on key issues, such as when integration is complete, and whether, or to what extent, policymakers' loyalties move to the supranational level. Haas himself later criticised the theory for its apparent inability to explain lulls in the EU integration process (Haas, 2001).

Neofunctionalist theorists consider integration to be a process, rather than agreeing on a specific outcome or end point. Haas defined integration as a 'process whereby political actors in several distinct national settings are persuaded to shift their loyalties, expectations and political activities towards a new centre, whose institutions possess or demand jurisdiction over the pre-existing national states' (Haas, 1958: 16). Emphasis is thus placed on the multiplicity of actors involved in EU decision-making processes. Furthermore, neofunctionalists regard the role and influence of supranational organisations and the development of transnational interest groups as important for deepening and expanding integration (Niemann and Schmitter, 2009).

Niemann and Schmitter (2009: 48–49) outline five assumptions of neofunctionalism. First, actors are rational and self-interested (Haas, 1970: 627), but also have the capacity to learn and change their preferences. Second, institutions can develop in a way that increases their autonomy from their creators. Third, incremental day-to-day decision making impacts the integration process to a greater extent than grand political strategies or designs. Fourth, supranational decision making is often about positive-sum games and compromises, rather than about veto-power. Fifth, interdependencies between economies tend to promote further integration.

Neofunctionalism's core concept is 'spillover', which is generally sub-divided into three types: 'political spillover', 'functional spillover' and 'cultivated spillover' (Niemann and Schmitter, 2009: 49–50). A process of spillover implies that decisions or actions taken in one particular policy sector lead to effects or pressures to move supranational policymaking to other policy areas (Pollack, 2010). This is described more specifically as functional spillover, referring to the organic process where cooperation in one policy area necessitates further cooperation in a (functionally) related area (Strøby-Jensen, 2007: 90). Political spillover describes deliberate political decisions taken to reach agreement in a variety of policy areas, and where supranational institutions (such as the Commission) and subnational actors/interest groups push for further integration.

Cultivated spillover describes the EU's supranational institution members' support for integration (Niemann and Schmitter, 2009: 50).

Two further concepts in neofunctionalist theory include 'elite socialisation' and the role of supranational interest groups. Elite socialisation describes the idea that the preferences and identities of people regularly involved in the policy-making process at, for example, the European level will gradually be 'Europeanised'. This implies a shift in loyalties from the national to the supranational level, as outlined in Haas' original definition of integration (Haas, 1958; Strøby-Jensen, 2007: 91). Also, neofunctionalism presents the idea that supranational interest groups develop over time and increase pressure for further integration (Strøby-Jensen, 2007: 92). Interest groups are also expected to become more 'European'.

Several theoretical criticisms have been levelled against neofunctionalism. First, critics have argued against the idea that actors in supranational institutions develop supranational loyalties, pointing instead to the example of EU member states' insistence on having their own national civil servants placed in EU institutions (Strøby-Jensen, 2007: 93). Second, neofunctionalism has been criticised for a lack of emphasis on the role of the member state as a driver or blocker of integration (Moravcsik, 1993). Third, the theory faced criticism over its normative aspirations as a model of integration, especially since it is regarded as 'elitist' and disregarding of the need for popular legitimacy (Risse, 2005: 297). Finally, Haas has criticised the theory himself due to its difficulties in describing the evident lack of progress in European integration at certain periods in history (Haas, 2001).

As neofunctionalism developed over time, scholars began to regard neofunctionalism not as a grand theory, but more as a theoretical perspective with partial explanatory power (Stone Sweet and Sandholtz, 1998). The 1990s saw a revival and a reformulation of neofunctionalism, ensuring its continued relevance. In conjunction with theories of institutionalism (see below), new scholarly work presented a version of neofunctionalist theory that emphasised transnational exchange and transactions (ibid.: 2). Thus, neofunctionalism could continue to demonstrate valuable explanatory power, especially when combined with other theoretical perspectives for a more complete picture of integration.

Liberal intergovernmentalism is another 'grand theory' of European integration. It aims to explain the large steps forward in integration (such as the major treaty revisions), rather than the incremental steps in the day-to-day running of the EU (Cini, 2007; Diez and Wiener, 2009; Moravcsik and Schimmelfennig, 2009). It is a state-centred theory of European integration that has been developed, in particular, by Andrew Moravcsik since the early 1990s (1993, 1998). Its intellectual roots are found in traditional 'intergovernmentalist' theory with 'rationalist institutionalist' assumptions (Hoffmann, 1966; Moravcsik and Schimmelfennig, 2009: 67). Moravcsik developed this theoretical account of integration in response to the perceived weaknesses in neofunctionalism, especially during those periods in the history of the EU when little progress on integration was made (Pollack, 2010). As an account of the European integration process, it is praised for its clarity and parsimony (Risse-Kappen, 1996: 63) and it remains one of the most influential of the European integration theories (Cini, 2007: 109). Moravcsik describes liberal intergovernmentalism as the 'baseline theory' of European and regional integration 'against which other theories are often compared' (Moravcsik and Schimmelfennig, 2009: 67).

Moravcsik summarises the main argument of liberal intergovernmentalism as 'European integration can best be explained as a series of rational choices made by national leaders' (Moravcsik, 1998: 18). Liberal intergovernmentalism is described as 'an application of "rationalist institutionalism"' (Moravcsik and Schimmelfennig, 2009: 67), so its assumptions are based on models of rational actor decision making. The main assumptions underlying this theoretical perspective are, therefore, that states are the key actors in the EU, and that they are rational, unitary actors. It is the member states' preferences that guide and shape European integration in general.

The theory follows a three-step model, whereby, first, member state preferences are formulated (at the national level, not at the supranational level as neofunctionalist theory might suggest); second, member states bargain at EU level; and, third, member states choose the institution to manage policy in a way that reflects member state commitments and recognition of the need for technical oversight (Moravcsik and Schimmelfennig, 2009; Pollack, 2010). Liberal intergovernmentalism is thus a state-centred theory, with these three factors (namely, state preferences; bargaining among states; and institutional choice) as the main explanatory variables to explain and understand advances in integration.

First, liberal intergovernmentalism aims to explain the motivations of states through an examination of their national preferences. Despite an acknowledgement that there is often a wide range of actors involved in policy formation, liberal intergovernmentalists still regard the state as a unitary actor since it is domestic politics and bargaining that generate the preferences of the state. Moravcsik also outlines that such preferences are subject to change over time, and that preferences vary among issue areas (Moravcsik and Schimmelfennig, 2009: 69).

Second, the interstate bargaining that takes place is explained following a rationalist perspective, and based on bargaining theory. In this context, liberal intergovernmentalists analyse the 'transaction costs' (or the costs of intergovernmental negotiating) involved (Moravcsik and Schimmelfennig, 2009: 71). Bargaining power may flow from several factors, including the relative need among the parties to ensure supranational agreement in a particular sector (and the balance of incentives/threats in the bargaining process), and the information at hand to the various member states on the details and implications of a particular policy issue.

Third, liberal intergovernmentalists seek to explain states' choice of institution for integration. In this respect, liberal intergovernmentalism follows on from 'neoliberal institutionalism' (Keohane and Nye, 1977). Here, the emphasis is on states' choice to delegate authority to supranational organisations in order to reduce future transaction costs, and to provide certainty about the preferences and behaviour of other states (Moravcsik and Schimmelfennig, 2009: 72).

There are four oft-cited criticisms of liberal intergovernmentalist theory in general. First, the theory is criticised for being too narrow in its scope to be considered a full-blown 'grand' theory of European integration. Specifically, liberal intergovernmentalism cannot sufficiently explain the day-to-day politics and functioning of the EU (Cini, 2007: 112), but focuses rather on explaining the great treaty reform decisions. Yet, empirically, scholars note that integration continues (although less visibly and perhaps less swiftly) through incremental steps in the daily operations of the EU.

A second criticism relates to Moravcsik's narrow conception of the state. Critics argue that liberal intergovernmentalism ignores or underestimates the subtleties of domestic politics. In order to arrive at an informed understanding of state preferences, critics argue, a more in-depth analysis of the component parts of the state, and of domestic politics, is required (Cini, 2007: 113). Member state preferences can evolve as a result of a myriad of factors, including economics, party ideologies, domestic civil society interest groups, and state institutions (Pollack, 2010: 20). The formulation and preservation of member state preferences is thus more complex than liberal intergovernmentalism admits.

Third, the theory is criticised for its lack of recognition of the role of supranational actors in the European integration process. This includes the roles of supranational institutions such as the Commission and the European Court of Justice, and also the role of non-state transnational actors or interest groups (Cini, 2007; Stone Sweet and Sandholtz, 1998: 12). This links to the constructivist criticism of liberal intergovernmentalism that ideas, norms and rules from the EU level necessarily impact the formulation of preferences at the domestic level, with interest groups at supra- and subnational levels interacting (Pollack, 2010: 21).

Finally, liberal intergovernmentalism is often criticised by proponents of historical institutionalism for its lack of emphasis on the unintended or consequential subsequent effects of decisions (Moravcsik and Schimmelfennig, 2009: 75). Historical institutionalists also lament the fact that liberal intergovernmentalism seems not to take into account the difficulty of reversing decisions, once institutions are in place (Pierson, 1998: 28–30; Pollack, 1998).

Moving on from liberal intergovernmentalism and neofunctionalism, then, are institutionalist-centred perspectives on European integration. Institutionalist analysis of the EU stems from a general re-insertion of institutions into many theoretical perspectives in international relations, including neorealism, Marxism and pluralism (Pollack, 2010). New institutionalist scholars contend that institutions 'matter' and should not be absent from theories of international relations, governance or regional integration (Hall and Taylor, 1996; March and Olsen, 1989). New institutionalist perspectives of European integration have developed since the 1980s and today can be divided into three versions: 'rational institutionalism', 'sociological institutionalism' and 'historical institutionalism' (Hall and Taylor, 1996; Pollack, 2009). These theoretical perspectives are, like neofunctionalism and unlike liberal intergovernmentalism, process-centred rather than state-centred theories (Jordan, Huitema, van Asselt, Rayner and Berkhout, 2010; Wiener and Diez, 2009a).

Rational institutionalism highlights the logic of consequentialism – actors behaving strategically to try to realise their own preferences (Risse, 2009: 147) – and transaction costs. Here, institutions are either the independent variable explaining how institutions shape outcomes, or else they are the dependent variable created and managed by rational actors to perform certain functions. Liberal intergovernmentalism took over the concepts and assumptions of rational institutionalism as part of its three-step model to explain European integration, where the institutional choice is a rational choice made by member states.

Sociological institutionalism describes how institutions 'constitute' actors and shape their views of the world, as individuals internalise the rules of behaviour (or culture) of institutions (Hall and Taylor, 1996: 948). Sociological institutionalism stemmed from sociology, rather than political science. Here, the emphasis is on a logic of appropriateness – i.e. actors' behaviour is guided by rules, and the desire to do 'the right thing' (Risse, 2009: 148), and thus 'individuals simultaneously constitute themselves as social actors ... and reinforce the convention to which they are adhering' (Hall and Taylor, 1996: 948). This is a constructivist understanding of the role of institutions (and the ideas, rules and norms associated with them) in preference formation and the shift of loyalty of actors working with and within institutions. Sociological institutionalism highlights the values and ideas that permeate institutions, rather than the rational objective of improved effectiveness, as an explanatory variable for the persistence or change in institutional set-up/scope.

Historical institutionalism, finally, focuses on the effects of institutions over time. In both historical and rational institutionalism, assumptions about the preferences of actors can be regarded as largely in line with rational choice theory (Pollack, 2009: 128). Rather than culture, socialisation, ideas and appropriateness of behaviour driving decisions, historical institutionalists consider the very structure of the institutional organisation as an important variable for explaining policymaking (Hall and Taylor, 1996: 937). The main concepts in historical institutionalism include the so-called 'lock-in' of decisions, and the fact that institutions are 'sticky' and cannot be easily reformed (Peters, 2011). Therefrom comes the idea of 'path dependence', where future decisions are taken in the context of past policy trends and ensuing unintended consequences of previous decisions (Hall and Taylor, 1996: 938). Institutions are regarded as playing a key role in politics and policymaking, and the effect and unintended consequences of past decisions ought not to be underestimated. Historical institutionalism also points to the existence of the 'joint-decision trap' (Scharpf, 1988, 2006), in which institutions remain in place, rigid and inflexible even in the face of a changing policy environment (Pierson, 1998: 45; Pollack, 2009).

The main criticism that is levelled against new institutionalist theories is that they are regarded as mid-level theories, and therefore cannot encompass or explain the causes of integration in general or on a grand scale. Instead, they can simply examine the effects of institutions on EU politics and policymaking (Pollack, 2009: 142). As such, it is only by linking the mid-level theory and analysis to a broader, grand theory that a fuller explanation of integration might be possible. Both historical and rational institutionalist perspectives often face a second criticism. Sociological institutionalists and constructivists question the basic assumptions underlying this institutionalist perspective about the rational nature of actors and institutions, which forms part of a greater debate between rationalist and constructivist scholars more generally (Pollack, 2010: 10). Such scholars argue in favour of taking account of the 'constitutive' effects of EU institutions on the individuals interacting with and within them (but see Dehousse and Thompson, 2012; Risse, 2009: 145).

Four variables to explain CPI

Each of the three theoretical perspectives discussed above emphasise particular elements or parts of EU integration, governance, policymaking and politics. Generally speaking, neofunctionalist scholars and institutionalist scholars consider that the European integration process can continue through further spillover among policy sectors and socialisation among European policymakers, elites and interest groups, and through 'sticky' institutional set-ups. Liberal intergovernmentalists, and also rationalist institutionalists, view the possibility of continued European integration more sceptically, drawing evidence on the limits or lack of socialisation from cases such as the difficulty faced in ratifying the Constitutional Treaty (Pollack, 2010). Taken together, these perspectives can help understand the many interacting elements at work in the EU policymaking process. The advantage of drawing on a 'mosaic' (Diez and Wiener, 2009: 19) of European integration theory for a study of CPI into the EU's energy sector is that no one variable is emphasised to the exclusion of others. In this section, I discuss how variables derived from the theories of European integration, when linked with previous empirical evidence describing critical explanatory variables in EPI literature, help develop an encompassing explanatory framework for the study of CPI in the EU. I describe four main explanatory variables for explaining and understanding CPI.

First, functional spillover, derived from neofunctionalism, can help explain the variation in the level of policy integration between policy areas according to the material interlinkages or 'functional interrelations' of the policy sectors in question. The concepts of cultivated and political spillover, and of elite socialisation can help explain the motivations of the member states and EU institutions – the Commission, Parliament and Council – for promoting (or blocking, ignoring) CPI. Neofunctionalism's emphasis on supranational interest groups and on the multiplicity of actors in the decision-making process also highlights the (lack of) pressure in a given policy sector for integrating climate policy objectives. Thus, from a neofunctionalist perspective, emphasis is placed on the role and influence of non-state actors in the decision-making process. For these reasons, a neofunctionalist perspective could help explain the extent of CPI with a particular emphasis on the functional interrelations of the issue areas and the involvement of stakeholders in the policy process. A heightened role for stakeholders in the policy process has often been regarded as an enabler of higher levels of EPI (EEA, 2005; Herodes, Adelle and Pallemaerts, 2007).

Second, liberal intergovernmentalism could also hold some potential for explaining the level of CPI in the EU with its emphasis on the role of member states. In particular, this perspective can shed light on the extent of political will and commitment of member states to CPI (and thus an understanding of the role of the Council and the European Council). The liberal intergovernmentalist emphasis on state actors, grand political decisions in the EU, and its focus on intergovernmental politics and member state preferences, point in the direction of grand political statements (from prime ministers, presidents and ministers) and their commitment to action. Different state preferences and their development over time can help us understand the level of political commitment at EU level and its evolution between 2000 and 2010.

Political commitment and leadership are generally considered very important for the establishment and development of policy integration (Jordan and Lenschow, 2008a; Lafferty and Hovden, 2003; Persson, 2004). Even if other enabling variables are evident, without an overarching commitment and leadership from the political level, it has been argued that policy integration will not succeed or be sustained (Jordan, 2002: 35). Later literature also suggests that political commitment is just one part of the puzzle. A recent state-of-the-art review of EPI found that political backing for EPI may exist in many jurisdictions (including in the EU), but without other enabling variables in place, this does not guarantee successful policy integration (Jordan and Lenschow, 2010: 147). Hence, a combination of explanatory variables is required for explaining levels of policy integration, of which one is political commitment.

Third, new institutionalist perspectives for the study of CPI can be useful especially due to their emphasis on policy pathways, and how institutional traditions and cultures impact the development of policy in the EU. The emphasis on how institutions matter in the policy process provides a complementary perspective to that presented in liberal intergovernmentalism and neofunctionalism. Thus, emphasis is placed on the importance of the EU's supranational institutions, their decision-making procedures (and/or traditions), and past policy decisions in policy development. Additionally, new institutionalist perspectives can complement liberal intergovernmentalism's focus on the political commitment of member states as an explanatory variable for CPI with an understanding that other institutions (the Parliament and Commission) could also demonstrate political commitment to CPI.

Both the institutional and policy context are highlighted as important variables for policy integration (EEA, 2005; OECD, 2002). This refers particularly to the governing architecture of the institutional framework for policymaking in the EU, and the influence of past policy and institutional decisions on present and future policymaking (Jordan and Lenschow, 2010; Pierson, 1998). For instance, we may hypothesise that an institutional set-up that requires unanimous decision making may lead to less CPI than decision making by qualified majority voting (QMV), as integrating climate policy objectives into other sectoral policies necessitates some change in policy, which can often be more easily achieved with majority voting rules. Where past policy actions have not achieved the desired results, this may allow for a window of opportunity to push for further policy developments. Furthermore, policymaking in the EU does not take place within a vacuum unconnected with external events, which may help push policy development.

In summary, the explanatory variables for a study of CPI in the EU that can be derived from theories of European integration and that are highlighted in empirical studies on EPI include: (1) functional interrelations; (2) political commitment; (3) institutional and policy context, each of which explain CPI in both the policy process and output; and (4) the process dimension to explain CPI in the policy output.

First, the nature of the *functional interrelations* between the two sectors being integrated can help explain the level of CPI found in the policy process (both at the policy proposal phase and with regard to amendments proposed throughout the policy negotiations) and in the policy output (in the text of the final policy output and in the specific aims of the final policy). As an explanatory variable, the nature of the functional interrelations between climate policy and the other policy sector in focus first forms the very basis of, and shapes any demand for, CPI. In the case of CPI in the EU's energy sector, the nature of the functional interrelations among the sectors describes the interaction between energy policy objectives and the objectives of climate policy. At the same time, the kind and strength of this functional demand may help in understanding the actual level of CPI. To this end, it is useful to consider two different properties of functional interrelations. First, functional interrelations may be more or less direct (interrelations between policy objectives are obvious, closely linked and clear) or indirect (interrelations between policy objectives may be more obscure, distant or hidden by other objectives), which may have repercussions for the strength of the resulting demand for CPI. Second, the functional interrelations may be more or less synergistic or conflictual, which may affect the ease or difficulty of advancing CPI. For example, where interrelations between policy objectives are obvious, such as when policies promoting renewable energy (RE) positively affect policies aiming to reduce GHG emissions, the functional interrelations are direct and synergistic. Policies to promote the construction of importing natural gas infrastructure interrelate with climate policies to the extent that increasing consumption of natural gas is counter to long-term climate objectives, yet may be beneficial (by displacing coal) in the short-term. Additionally, such policies may divert funds and attention away from RE policies, and may focus on security of supply questions. In this second case, the functional interrelations are less direct, and are mainly conflictual.

Table 2.3 outlines a matrix of functional interrelations in relation to CPI. Where the functional interrelations between climate policy and the policy sector under development are direct and synergistic, higher levels of CPI could be expected, while indirect and conflictual functional interrelations could more likely block CPI from occurring. Thus, we can describe situations where the functional interrelations between a policy sector and long-term climate policy objectives are both direct and synergistic as providing *most favourable*

Table 2.3 The nature of functional interrelations and their potential effect on CPI

	Direct	Indirect
Synergistic	+ +	+ −
Conflictual	− +	− −

conditions for the advancement of CPI in the policy process and in the policy output.

Second, the *political commitment* demonstrated by the EU's leaders (including, but not only, heads of state and government) could be considered a core variable for explaining CPI in both the policy process and output. Evidence for political commitment will surface in the conclusions of the Council and the European Council. New institutionalist perspectives direct us also to consider the political commitment demonstrated by other EU institutions – thus the policy proposals and statements of the Commission and the proposed amendments and statements of the Parliament may also provide an indication of the level of political commitment overall in the EU. Political commitment is thus recognised in the stated objectives of the political members of the EU, and in follow-up, in the decisions taken to see these objectives materialise.

With regard to CPI, two aspects of political commitment appear to be particularly relevant. First, a general political commitment to combating climate change plays an important overarching role when it comes to CPI. Second, the political commitment to climate policy *integration* into the specific policy sector under investigation, based on an acknowledgement of the relevance of any overall climate policy objectives for the sector, should be important for the prospects of CPI actually being advanced. Political commitment to climate policy and CPI from the outset of policymaking can affect the extent to which the functional interrelations between the policy sector and climate policy are emphasised in the policy proposal and throughout the policy negotiations. High levels of political commitment among EU policymakers may push CPI to the extent that the final result is a policy output that is far-reaching in terms of achieving climate policy objectives. The political commitment can thus explain the level of CPI in the policy output as well as in the policy process.

Here, I measure political commitment to climate policy and to CPI with reference to the public statements of EU, and assess the existence of concrete follow-up on the expressed commitment (in terms of legislation, for example). Table 2.4 outlines the scale for measuring political commitment. I use a threefold scale in this case, as assessing the EU's political commitment to combating climate change and to advancing CPI is necessarily a highly qualitative task. The indicators used to identify the different levels of CPI vary according to the evidence of commitment in political statements, followed up with concrete action.

Third, a new institutionalist perspective leads us to pay particular attention to the *institutional and policy context* for policy integration in both the policy process and the policy output. The institutional context refers, particularly, to the EU's

Table 2.4 Measuring political commitment to, first, combating climate change, and, second, advancing CPI

	Low	Medium	High
Measuring political commitment	No evidence of commitment to climate change or CPI in statements	Expressed commitment to climate change or CPI in statements	Expressed commitment to climate change or CPI in statements, and actions to follow through on the commitment

institutional set-up, including the decision-making procedures in the institutions. The policy context can refer to the path dependency that is created for present and future policies, due to past policy and institutional decisions (Jordan and Lenschow, 2010; Pierson, 1998), as well as external policy factors that may influence policy development in the EU. We could expect certain external 'shocks' (such as energy or financial crises) or events (such as international climate negotiations) to impact the policymaking process (and policy priorities) within the EU, by opening windows of opportunity for policy development (Kingdon, 2003; Nohrstedt, 2006; Wettestad, 2005). Actors may learn from past policy failures or successes and previous decisions may create or undermine a dynamic that facilitates a push for policy change throughout the policy process. Decision-making procedures may also affect the extent of CPI in the policy output. Decisions by qualified majority, for example, may be assumed to facilitate policy change towards CPI as compared with a unanimity rule.

To assess this variable I analyse policy documents to discover the institutional set-up in the policy sector in question. The particular policymaking procedure is relevant as it changes the decision making rule (and thus whether QMV rules apply). I look to the policy context, which involves assessing the timing and sequencing of the policy negotiations, and whether or not the external political context can be considered to play a role in policy development. This variable explains levels of CPI in that the more emphasis on combating climate change, both in the institutional context and the external political context, the greater the expectations of finding higher levels of CPI. Thus, the institutional and policy contexts can be said to be either more or less favourable for CPI in both the policy process and the policy output, and especially with regard to the levels of CPI in the policy process, although certain of these contextual issues may swing the balance in the final policy decision.

Finally, I expect the level of CPI in the policy process to play an explanatory role for understanding the extent of CPI in the policy output. The higher the level of CPI in the policy process, the more likely higher levels of CPI will also be found in the policy output (Briassoulis, 2005c). This explanatory variable relates to both new institutionalist and neofunctionalist theoretical perspectives, which emphasise the importance of multiple actors in the policymaking processes. Whether procedures are in place in the policy process to allow access for environmental and climate actors can affect the level of CPI in the policy process, which in turn may explain the extent of CPI in the policy output. One assumption here is that where CPI does not feature highly in the policy process it would be surprising (but not entirely impossible) to then find high levels of CPI in the policy output.

Furthermore, for functional interrelations and their properties to have political effect, they need to be part of the political discourse and realised in the policy process. This final aspect also requires attention in the analysis. For the purposes of a study on CPI, an analysis of the functional interrelations of policy sectors under examination can be usefully complemented with a document analysis, and with interviewee responses, to establish whether it is acknowledged in the various sectors that such functional interrelations exist and can affect the policy objectives of that sector. Thus, breaking down the extent of CPI in the policy process could help provide more understanding of the level of CPI in the policy output. The overall level of CPI in the policy process, as measured according to the indicators discussed above, therefore, can affect the overall level of CPI in the policy output.

It is important to note that these explanatory variables are unlikely to explain the levels of CPI by themselves, but interact with each other to provide a clearer explanation or understanding of CPI. While some may prove more important in the explanation than others, I nevertheless expect the combination of these variables to provide a more nuanced understanding of why certain levels of CPI are found. A first hypothesis about the hierarchy, importance or weight of these variables, based on the literature on EPI, is that political commitment to climate policy and to climate policy integration will prove crucial for the emergence of CPI in EU energy policy. However, I would suggest that political commitment is an insufficient explanatory variable. It cannot, by itself, explain levels of CPI in EU energy policy process and output. A more nuanced hypothesis would suggest that a combination of variables is required for CPI to manifest itself in the policy process and output, of which political commitment is one important one (Adelle and Russel, 2013; Jordan and Lenschow, 2010). Furthermore, we could hypothesise that it is in fact the recognition of functional interrelations that is a baseline and crucial condition for CPI. Where a policy sector interrelates directly and synergistically with climate policy, political commitment to CPI and high levels of CPI in the policy process could more likely be manifested. The opposite (indirect and conflictual functional interrelations combined with low political commitment and low CPI in the policy process) may also occur. Therefore, we can consider that functional interrelations must crucially be recognised in the policy process. Following from a low level of CPI in the policy process, low levels of CPI in the output may also be expected. The remaining explanatory variables, such as the institutional and policy context, can shift the balance of CPI to lower or higher levels, but perhaps only if the functional interrelations had already been recognised from the early stages of the policy process. The external policy context and the institutional context can be said, thus, to represent background contextual factors that can interact with any of the remaining variables in such a way that they can have more or less effect on CPI. Therefore, the four explanatory variables are not expected to hold equal weight in all case studies, and their relative importance may depend on the existence, first and foremost, of the recognition of the functional interrelations.

Summary

This chapter outlined the conceptual and analytical frameworks applied to the empirical study of CPI into the EU's energy sector. Conceptualising CPI as a matter of degree, where 'principled priority' for climate objectives is considered a methodological tool for measuring very high levels of CPI provides the advantage that the analysis can encompass all levels of CPI, from weak to strong, and enhance comparison with other similar studies. Operationalising this concept requires the measurement of CPI in both the policy output and the policy process. With regard to the policy output, a very high level of CPI would mean that the policy sector objectives are very close to 100 per cent in line with (scientifically-grounded) climate objectives – such as the long-term objective to reduce GHG emissions in the EU by between 80 and 95 per cent by 2050 (based on IPCC recommendations for developed country action to ensure that global temperature increase does not exceed 2° Celsius). For the policy process, procedures guaranteeing the access to the policymaking process for internal (within the EU institutions) and external pro-climate policy stakeholders, as well as the recognition and articulation of the functional interrelations with long-term climate policy objectives in the policy process, can enable higher levels of CPI.

Overall, the review of literature on EPI and of general theories of European integration allows for the identification of a limited number of core explanatory variables. These can be systematically employed to explain varying levels of policy integration in general, and CPI in particular, across a number of cases. Taken together, these core variables and their identified components provide a differentiated but manageable framework for the systematic exploration and explanation of CPI into other policies. This explanatory framework highlights four main explanatory variables: the nature of *functional interrelations*; *political commitment* to climate policy and to CPI; and the *institutional and policy context*, which each explain CPI in both the policy process and output; and, the *process dimension*, which can explain CPI in the policy output only. For a summary of the explanatory framework, see Table 2.5.

The next three chapters measure the extent of CPI in three cases of energy policy in the EU. Chapter 3 examines CPI in the EU's renewable energy policy, Chapter 4 explores CPI in the EU's energy performance of buildings policies, and Chapter 5 analyses CPI in the EU's policies to support natural gas import infrastructure. Each case is examined over time, with a focus on the period 2000 to 2010. Chapter 6 applies the explanatory framework to the results and findings of the three cases.

Table 2.5 Summary of explanatory framework

Variable	Theoretical perspectives	Explains CPI in process or output?	Elements	Operationalisation
Functional interrelations	Neofunctionalism	Process and output	Direct or indirect; synergetic or conflictual	Nature and type of functional interrelations between two sectoral policy objectives
Political commitment	Liberal intergovernmentalism and new institutionalism	Process and output	Overarching to climate policy objectives; to CPI, in particular	Council and European Council conclusions; statements and follow-up of EU institutions (including Com and EP)
Institutional and policy context	New institutionalism and neofunctionalism	Process and output	Institutional context (procedures); past policy decisions; external shocks or events	Policymaking procedure; timing of policy process
Process dimension	Neofunctionalism and new institutionalism	Output only	Extent of CPI in process; discourse on climate policy in process	Pro-climate stakeholder and advocate involvement and access to the policy process; acknowledgement of functional interrelations

References

Adelle, C. and Russel, D. (2013). Climate Policy Integration: a Case of Déjà Vu? *Environmental Policy and Governance*, 23(1), 1–12.

Ahmed, I. H. (2009). *Climate Policy Integration: Towards Operationalization*. DESA Working Paper no. 73: ST/ESA/DWP/73. New York: UN/DESA.

Biermann, F., Davies, O. and van der Grip, N. (2009). Environmental Policy Integration and the Architecture of Global Environmental Governance. *International Environmental Agreements: Politics, Law and Economics*, 9(4), 351–369.

Briassoulis, H. (2005a). Analysis of Policy Integration: Conceptual and Methodological Considerations. In H. Briassoulis (ed.), *Policy Integration for Complex Environmental Problems: the Example of Mediterranean Desertification* (pp. 50–80). Aldershot: Ashgate.

Briassoulis, H. (2005b). Complex Environmental Problems and the Quest for Policy Integration. In H. Briassoulis (ed.), *Policy Integration for Complex Environmental Problems: the Example of Mediterranean Desertification* (pp. 1–49). Aldershot: Ashgate.

Briassoulis, H. (ed.) (2005c). *Policy Integration for Complex Environmental Problems: the Example of Mediterranean Desertification*. Aldershot: Ashgate.

Bryner, G. C. (2012). *Integrating Climate, Energy and Air Pollution Policies* (with Robert J. Duffy). Cambridge, MA: MIT Press.

Cini, M. (2007). Intergovernmentalism. In M. Cini (ed.), *European Union Politics*, 2nd edn. (pp. 99–116). Oxford: Oxford University Press.

Dehousse, R. and Thompson, A. (2012). Intergovernmentalists in the Commission: Foxes in the Henhouse? *European Integration*, 34(2), 113–132.

Diez, T. and Wiener, A. (2009). Introducing the Mosaic of Integration Theory. In A. Wiener and T. Diez (eds), *European Integration Theory*, 2nd edn. (pp. 1–22). Oxford: Oxford University Press.

Dupont, C. (2011). Climate Policy Integration in the EU. In W. L. Filho (ed.), *The Economic, Social and Political Elements of Climate Change* (pp. 385–404). Berlin: Springer-Verlag.

Dupont, C. and Oberthür, S. (2012). Insufficient Climate Policy Integration in EU Energy Policy: the Importance of the Long-Term Perspective. *Journal of Contemporary European Research*, 8(2), 228–247.

Dupont, C. and Oberthür, S. (2015a). Decarbonization in the EU: Setting the Scene. In C. Dupont and S. Oberthür (eds), *Decarbonization in the European Union: Internal Policies and External Strategies* (pp. 1–24). Houndmills: Palgrave Macmillan.

Dupont, C. and Oberthür, S. (eds). (2015b). *Decarbonization in the European Union: Internal Policies and External Strategies*. Houndmills: Palgrave Macmillan.

Dupont, C. and Primova, R. (2011). Combating Complexity: the Integration of EU Climate and Energy Policies. *European Integration Online Papers*, 15(Special mini-issue 1), Article 8.

ECF. (2010). *Roadmap 2050. A Practical Guide to a Prosperous, Low Carbon Europe*. Brussels: European Climate Foundation.

EEA. (2005). *Environmental Policy Integration in Europe: State of Play and Evaluation Framework. EEA Report no 2/2005*. Copenhagen: European Environment Agency.

EREC. (2010). *RE-thinking 2050: a 100% Renewable Energy Vision for the European Union*. Brussels: European Renewable Energy Council.

EREC and Greenpeace. (2010). *Energy [R]evolution: Towards a Fully Renewable Energy Supply in the EU 27*. Brussels: Greenpeace International and European Renewable Energy Council.

Eurelectric. (2010). *Power Choices. Pathways to Carbon-Neutral Electricity in Europe by 2050*. Brussels: Eurelectric.

Eurogas. (2011). *Eurogas Roadmap 2050*. Brussels: Eurogas.

European Commission. (1973). Programme of Environmental Action of the European Communities. Part II: Detailed Description of the Actions to be Undertaken at Community Level Over the Next Two Years. COM(73) 530.

European Commission. (2011). Communication from the Commission: Energy Roadmap 2050. COM(2011) 885/2.

European Council. (2009). *Presidency Conclusions*, October 2009. Brussels: Council of the European Union.

Friedrichs, J. and Kratochwil, F. (2009). On Acting and Knowing: How Pragmatism Can Advance International Relations Research and Methodology. *International Organization*, 63(4), 701–731.

Haas, E. B. (1958). *The Uniting of Europe: Political, Social and Economic Forces, 1950–1957.* Stanford, CA: Stanford University Press.

Haas, E. B. (1970). The Study of Regional Integration: Reflections on the Joy and Anguish of Pretheorizing. *International Organization,* 24(4), 607–646.

Haas, E. B. (2001). Does Constructivism Subsume Neofunctionalism? In T. Christiansen, K. E. Jørgensen and A. Wiener (eds), *The Social Construction of Europe* (pp. 22–31). London: Sage.

Hall, P. A. and Taylor, R. C. R. (1996). Political Science and the Three New Institutionalisms. *Political Studies,* 44(5), 936–957.

Hauser, H. (2011). European Union Lobbying Post Lisbon: an Economic Analysis. *Berkeley Journal of International Law,* 29(2), 680–709.

Heaps, C., Erickson, P., Kartha, S. and Kemp-Benedict, E. (2009). *Europe's Share of the Climate Challenge: Domestic Actions and International Obligations to Protect the Planet.* Stockholm: Stockholm Environment Institute.

Herodes, M., Adelle, C. and Pallemaerts, M. (2007). *Environmental Policy Integration at the EU Level: a Literature Review.* EPIGOV Paper No. 5. Berlin: Ecologic, Institute for International and European Environmental Policy.

Hoffmann, S. (1966). Obstinate of Obsolete? The Fate of the Nation-State and the Case of Western Europe. *Daedalus,* 95(5), 862–915.

IEA. (2011a). *World Energy Outlook 2011.* Paris: OECD/International Energy Agency.

IEA. (2011b). *World Energy Outlook 2011: Are We Entering a Golden Age of Gas?* Paris: OECD/International Energy Agency.

IPCC. (2007). *Climate Change 2007. Fourth Assessment Report: Synthesis Report.* Geneva: Intergovernmental Panel on Climate Change.

IPCC. (2013). Summary for Policymakers. In T. F. Stoker, D. Qin, G.-K. Plattner, M. Tignor, S. K. Allen, J. Boschung, ... P. M. Midgley (eds), *Climate Change 2013: the Physical Science Basis. Contribution of Working Group I to the Fifth Assessment Report of the Intergovernmental Panel on Climate Change.* Cambridge: Cambridge University Press.

Jacob, K., Volkery, A. and Lenschow, A. (2008). Instruments for Environmental Policy Integration in 30 OECD Countries. In A. Jordan and A. Lenschow (eds), *Innovation in Environmental Policy? Integrating the Environment for Sustainability* (pp. 24–45). Cheltenham: Edward Elgar.

Jordan, A. (2002). Efficient Hardware and Light Green Software: Environmental Policy Integration in the UK. In A. Lenschow (ed.), *Environmental Policy Integration: Greening Sectoral Policies in Europe* (pp. 35–56). London: Earthscan.

Jordan, A., Huitema, D., van Asselt, H., Rayner, T. and Berkhout, F. (2010). *Climate Change Policy in the European Union: Confronting the Dilemmas of Mitigation and Adaptation?* Cambridge: Cambridge University Press.

Jordan, A. and Lenschow, A. (eds) (2008a). *Innovation in Environmental Policy? Integrating the Environment for Sustainability.* Cheltenham: Edward Elgar.

Jordan, A. and Lenschow, A. (2008b). Integrating the Environment for Sustainable Development: an Introduction. In A. Jordan and A. Lenschow (eds), *Innovation in Environmental Policy? Integrating the Environment for Sustainability* (pp. 3–23). Cheltenham: Edward Elgar.

Jordan, A. and Lenschow, A. (2010). Environmental Policy Integration: a State of the Art Review. *Environmental Policy and Governance,* 20(3), 147–158.

Jordan, A., Schout, A. and Unfried, M. (2008). The European Union. In A. Jordan and A. Lenschow (eds), *Innovation in Environmental Policy? Integrating the Environment for Sustainability* (pp. 159–179). Cheltenham: Edward Elgar.

Keohane, R. O. and Nye, J. (1977). *Power and Independence: World Politics in Transition*. Boston, MA: Little, Brown and Company.
Kingdon, J. W. (2003). *Agendas, Alternatives, and Public Policies*, 2nd edn. London: Longman.
Kulovesi, K., Morgera, E. and Muñoz, M. (2010). The EU's Climate and Energy Package: Environmental Integration and International Dimensions. *Edinburgh Europa Paper Series*, 2010(38).
Lafferty, W. M. and Hovden, E. (2003). Environmental Policy Integration: Towards an Analytical Framework. *Environmental Politics*, 12(5), 1–22.
Lafferty, W. M. and Knudsen, J. (2007). *The Issue of 'Balance' and Trade-Offs in Environmental Policy Integration: How Will We Know EPI When We See It? EPIGOV Paper No. 11*. Berlin: Ecologic, Institute for International and European Environmental Policy.
Lechtenböhmer, S. and Samadi, S. (2015). The Power Sector: Pioneer and Workhorse of Decarbonization. In C. Dupont and S. Oberthür (eds), *Decarbonization in the European Union: Internal Policies and External Strategies* (pp. 46–69). Houndmills: Palgrave Macmillan.
Lenschow, A. (ed.) (2002a). *Environmental Policy Integration: Greening Sectoral Policies in Europe*. London: Earthscan.
Lenschow, A. (2002b). Greening the European Union: an Introduction. In A. Lenschow (ed.), *Environmental Policy Integration: Greening Sectoral Policies in Europe* (pp. 3–21). London: Earthscan.
Lenschow, A. (2010). Environmental Policy: Contending Dynamics of Policy Change. In H. Wallace, M. A. Pollack and A. R. Young (eds), *Policy-Making in the European Union*, 6th edn. (pp. 307–330). Oxford: Oxford University Press.
Lindberg, L. N. (1963). *The Political Dynamics of European Economic Integration*. Stanford, CA: Stanford University Press.
March, J. G. and Olsen, J. P. (1989). *Rediscovering Institutions: the Organizational Basis of Politics*. New York: The Free Press.
Metcalfe, L. (1994). International Policy Co-ordination and Public Management Reform. *International Review of Administrative Sciences*, 60(2), 271–290.
Mickwitz, P., Beck, S., Jensen, A., Pedersen, A. B., Aix, F., Carss, D., ... Van Bommel, S. (2009). Climate Policy Integration as a Necessity for an Efficient Climate Policy. Paper presented at *Human Dimensions of Global Environmental Change Conference*, Amsterdam.
Moravcsik, A. (1993). Preferences and Power in the European Community: a Liberal Intergovernmentalist Approach. *Journal of Common Market Studies*, 31(4), 473–524.
Moravcsik, A. (1998). *The Choice for Europe. Social Purpose and State Power from Messina to Maastricht*. London: Routledge/UCL Press.
Moravcsik, A. and Schimmelfennig, F. (2009). Liberal Intergovernmentalism. In A. Wiener and T. Diez (eds), *European Integration Theory*, 2nd edn. (pp. 67–87). Oxford: Oxford University Press.
Niemann, A. and Schmitter, P. C. (2009). Neofunctionalism. In A. Wiener and T. Diez (eds), *European Integration Theory*, 2nd edn. (pp. 45–66). Oxford: Oxford University Press.
Nilsson, M., Eckerberg, K., Hagberg, L., Swartling, Å. G. and Söderberg, C. (2007). Policy Framing and EPI in Energy and Agriculture. In M. Nilsson and K. Eckerberg (eds), *Environmental Policy Integration in Practice: Shaping Institutions for Learning* (pp. 85–110). London: Earthscan.

Nilsson, M. and Persson, Å. (2003). Framework for Analysing Environmental Policy Integration. *Journal of Environmental Policy and Planning*, 5(4), 333–359.

Nohrstedt, D. (2006). External Shocks and Policy Change: Three Mile Island and Swedish Nuclear Energy Policy. *Journal of European Public Policy*, 12(6), 1041–1059.

Nollkamper, A. (2002). Three Conceptions of the Integration Principle in International Environmental Law. In A. Lenschow (ed.), *Environmental Policy Integration: Greening Sectoral Policies in Europe* (pp. 22–34). London: Earthscan.

Oberthür, S. (2009). Interplay Management: Enhancing Environmental Policy Integration Among International Institutions. *International Environmental Agreements: Politics, Law and Economics*, 9(4), 371–391.

OECD. (2002). *Improving Policy Coherence and Integration for Sustainable Development: A Checklist*. Paris: Organisation for Economic Co-operation and Development.

Persson, Å. (2004). *Environmental Policy Integration: An Introduction. PINTS – Policy Integration for Sustainability Background Paper*. Stockholm: Stockholm Environment Institute.

Persson, Å. (2007). Different Perspectives on EPI. In M. Nilsson and K. Eckerberg (eds), *Environmental Policy Integration in Practice: Shaping Institutions for Learning* (pp. 25–48). London: Earthscan.

Peters, B. G. (1998). Managing Horizontal Government: the Politics of Co-ordination. *Public Administration*, 76(2), 295–311.

Peters, B. G. (2011). *Institutional Theory in Political Science: the New Institutionalism*, 3rd edn. London: Continuum Publishing Corporation.

Pierson, P. (1998). The Path to European Integration: a Historical-Institutionalist Analysis. In W. Sandholtz and A. Stone Sweet (eds), *European Integration and Supranational Governance* (pp. 27–58). Oxford: Oxford University Press.

Pollack, M. A. (1998). The Engines of Integration? Supranational Autonomy and Influence in the European Union. In W. Sandholtz and A. Stone Sweet (eds), *European Integration and Supranational Governance* (pp. 217–249). Oxford: Oxford University Press.

Pollack, M. A. (2009). The New Institutionalisms and European Integration. In A. Wiener and T. Diez (eds), *European Integration Theory*, 2nd edn. (pp. 125–143). Oxford: Oxford University Press.

Pollack, M. A. (2010). Theorizing EU Policy-Making. In H. Wallace, M. A. Pollack and A. R. Young (eds), *Policy-Making in the European Union*, 6th edn. (pp. 15–44). Oxford: Oxford University Press.

Risse, T. (2005). Neofunctionalism, European Identity, and the Puzzles of European Integration. *Journal of European Public Policy*, 12(2), 291–309.

Risse, T. (2009). Social Constructivism and European Integration. In A. Wiener and T. Diez (eds), *European Integration Theory*, 2nd edn. (pp. 144–160). Oxford: Oxford University Press.

Risse-Kappen, T. (1996). Exploring the Nature of the Beast: International Relations Theory and Comparative Policy Analysis Meet the European Union. *Journal of Common Market Studies*, 34(1), 53–80.

Rosamond, B. (2007). New Theories of European Integration. In M. Cini (ed.), *European Union Politics*, 2nd edn. (pp. 117–136). Oxford: Oxford University Press.

Sabatier, P. A. (2007). The Need for Better Theories. In P. A. Sabatier (ed.), *Theories of the Policy Process*, 2nd edn. (pp. 3–17). Boulder, CO: Westview Press.

Scharpf, F. W. (1988). The Joint-Decision Trap: Lessons from German Federalism and European Integration. *Public Administration*, 66(3), 239–278.

Scharpf, F. W. (2006). The Joint-Decision Trap Revisited. *Journal of Common Market Studies*, 44(4), 845–864.

Schout, A. and Jordan, A. (2008). The European Union's Governance Ambitions and its Administrative Capacities. *Journal of European Public Policy*, 15(7), 957–974.

Selianko, I. and Lenschow, A. (2015). Energy Policy Coherence from an Intra-Institutional Perspective: Energy Security and Environmental Policy Coordination Within the European Commission. *European Integration Online Papers*, 19(Special issue 1), Article 2.

Stone Sweet, A. and Sandholtz, W. (1998). Integration, Supranational Governance, and the Institutionalization of the European Polity. In W. Sandholtz and A. Stone Sweet (eds), *European Integration and Supranational Governance* (pp. 1–26). Oxford: Oxford University Press.

Strøby-Jensen, C. (2007). Neo-functionalism. In M. Cini (ed.), *European Union Politics*, 2nd edn. (pp. 85–98). Oxford: Oxford University Press.

Underdal, A. (1980). Integrated Marine Policy: What? Why? How? *Marine Policy*, 4(3), 159–169.

Van Bommel, S. and Kuindersma, W. (2008). *Policy Integration, Coherence and Governance in Dutch Climate Policy: a Multi-Level Analysis of Mitigation and Adaptation Policy*. Alterra-rapport 1799. Wageningen: Alterra.

Vasileiadou, E. and Tuinstra, W. (2013). Stakeholder Consultations: Mainstreaming Climate Policy in the Energy Directorate? *Environmental Politics*, 22(3), 475–495.

Von Homeyer, I. (2006). Environmental Policy Integration and Modes of Governance – State-of-the-Art Report. *EPIGOV Paper No. 2*. Berlin: Ecologic, Institute for International and European Environmental Policy.

Warleigh-Lack, A. and Drachenberg, R. (2010). Policy Making in the European Union. In M. Cini and N. Pérez-Solórzano Borragán (eds), *European Union Politics*, 3rd edn. (pp. 209–224). Oxford: Oxford University Press.

Wettestad, J. (2005). The Making of the 2003 EU Emissions Trading Directive: An Ultra-Quick Process due to Entrepreneurial Proficiency? *Global Environmental Politics*, 5(1), 1–23.

Wiener, A. and Diez, T. (eds) (2009a). *European Integration Theory*. Oxford: Oxford University Press.

Wiener, A. and Diez, T. (2009b). Taking Stock of Integration Theory. In A. Wiener and T. Diez (eds), *European Integration Theory*, 2nd edn. (pp. 241–252). Oxford: Oxford University Press.

Wilkinson, D. (1998). Steps Towards Integrating the Environment into other EU Policy Sectors. In T. O'Riordan and H. Voisey (eds), *The Transition to Sustainability: the Politics of Agenda 21 in Europe* (pp. 113–129). London: Earthscan.

Wilkinson, D. (2007). *Environmental Policy Integration at EU Level – State-of-the-Art Report*. EPIGOV Paper No. 4. Berlin: Ecologic, Institute for International and European Environmental Policy.

World Commission on Environment and Development. (1987). *Our Common Future*. WCED.

WWF. (2011). *The Energy Report: 100% Renewable Energy by 2050*. Gland, Switzerland: World Wide Fund for Nature.

Zingerli, C., Bisang, K. and Zimmermann, W. (2004). Towards Policy Integration: Experiences with intersectoral coordination in international and national forest policy. *Berlin Conference 2004 on the Human Dimension of Global Environmental Change 'Greening of Policies – Interlinkages and Policy Integration.'* Berlin.

3 EU policy on renewable energy

The empirical analysis opens in this chapter with an examination of the levels of climate policy integration in the policy process and output of EU renewable energy (RE) policy, while also exploring the development of EU RE policy over time. I analyse the 2001 renewable electricity Directive (RES-E Directive 2001/77/EC) and the 2009 renewable energy Directive (RE Directive 2009/28/EC). I begin with a discussion of RE in the EU and historical policy developments to promote RE. I then describe the RES-E and RE Directives and measure CPI in their processes and outputs. The chapter closes with a brief discussion of the future outlook for CPI in RE policy.

Renewable energy in the EU

Renewable sources of energy are defined as renewable non-fossil energy sources that are 'replenished by natural sources at a rate that equals or exceeds' their rate of use (Moomaw *et al.*, 2011: 3). These sources are identified in EU legislation as including wind, solar, geothermal, wave, tidal, hydropower, biomass, landfill gas, sewage treatment plant gas and biogases (Directive 2001/77/EC Article 2.a; Directive 2009/28/EC Article 2.a). According to data from the European Environment Agency (EEA) and Eurostat, the share of RE in the EU28's final energy consumption increased from about 8.3 per cent in 2004 to 15 per cent in 2013 (EEA, 2014; Eurostat, 2015). Clearly, progress had been made over the two decades from 1990, when the RE share stood at about 4–6 per cent (see Figure 3.1; EEA, 2011; European Commission, 1992). The main increases in the share of RE happened rather in the period from 2005 onwards, when stronger policy instruments were in place and with the effect of the economic and financial crises from 2008 onwards leading to an artificial inflation of the overall 'share' of RE in final energy consumption (as consumption dropped) (EEA, 2011).

Although the share of RE has increased in the EU, this has not happened uniformly across all member states. In 2013, Sweden had the greatest share of RE in its final energy consumption (over 52 per cent, up from nearly 39 per cent in 2004), while Luxembourg had an estimated 3.6 per cent share of RE in its final energy consumption in 2013 (compared to 0.9 per cent in 2004; see Table 3.1 for details of member state RE development). Nevertheless, each member state has

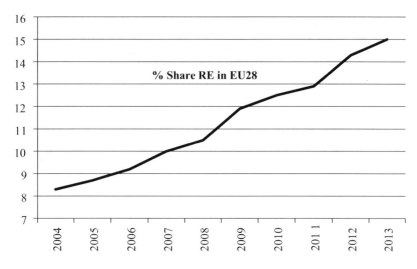

Figure 3.1 Percentage share of renewable energy in final energy consumption in the EU28

Source: Compiled from (EEA, 2014; Eurostat, 2015)

increased its share of RE over time, and the EU as a whole is on track to achieve an overarching target of at least 20 per cent share of RE in final energy consumption in 2020 (EEA, 2014).

Early policy development

While the increases of the share of RE in the EU is evident especially since 2005, attempts at EU-wide RE policy development began much earlier. The EU began discussing RE policy in the 1980s, based on multiple motivations: responding to environmental and climate concerns, tackling energy security issues and developing an internal energy market.

In the wake of the oil crises in the 1970s, following the Arab-Israeli war and the Iranian revolution, some EU member states began to develop policies to promote RE. These crises imbued a sense of energy security vulnerability. With few EU member states having sufficient proven reserves of fossil fuels, states began to search for alternative energy sources that could increase energy independence (Hildingsson, Stripple and Jordan, 2010: 105). These early policy developments took place at the level of some member states, and not at the EU level. Denmark, Sweden and Germany, in particular, first developed national RE policies in the 1970s (Nilsson, 2011: 113). By the 1980s, several member states had put in place their own instruments and support mechanisms for RE, resulting in several different approaches across the EU (Lauber, 2001). The member

Table 3.1 Percentage share of RE in final energy consumption in the EU28 and in each member state in 2004 and 2013, compared to 2020 target

	2004	2013	2020 target
EU28	7.9	15	20
Austria	22.7	32.6	34
Belgium	1.9	7.9	13
Bulgaria	9.5	19	16
Croatia	13.2	18	20
Cyprus	3.1	8.1	13
Czech Republic	5.9	12.4	13
Denmark	14.5	27.2	30
Estonia	18.4	25.6	25
Finland	29.2	36.8	38
France	9.4	14.2	23
Germany	5.8	12.4	18
Greece	6.9	15	18
Hungary	4.4	9.8	13
Ireland	2.4	7.8	16
Italy	5.6	16.7	17
Latvia	32.8	37.1	40
Lithuania	17.2	23	23
Luxembourg	0.9	3.6*	11
Malta	0.1	3.8	10
Netherlands	1.9	4.5	14
Poland	6.9	11.3	15
Portugal	19.2	25.7	31
Romania	17	23.9	24
Slovenia	16.1	21.5	25
Slovakia	5.7	9.8	14
Spain	8.3	15.4	20
Sweden	38.7	52.1	49
United Kingdom	1.2	5.1	15

Note: * the figure for Luxembourg for 2013 is estimated.

Source: Compiled from (Eurostat, 2015)

states also favoured different renewable technologies (Denmark supported wind, Sweden favoured bioenergy crops, for example, Nilsson, 2011: 113).

Environmental concerns did begin to rise up the European policy agenda in the 1980s and 1990s (Lenschow, 2002), along with international attention to environmental issues. The problem of climate change began to drive RE policy in the EU, as an add-on to the energy security concerns highlighted during the 1970s. EU policy aimed to improve coordination in achieving an overall increase in the share of renewables in the energy mix of the EU, thus achieving both energy security and environmental objectives. EU RE policy, however, also emerged in response to the push to develop a liberalised internal energy market (Hildingsson *et al.*, 2010: 103 and 106). With the drive towards market liberalisation, in the late 1980s and 1990s, issues arose related to the perceived unequal playing field among RE producers and conventional energy producers (Jansen and Uyterlinde, 2004: 93). The Commission, especially, identified a need for further liberalisation and for the development of EU-level harmonised RE policies (Boasson and Wettestad, 2013).

The Council outlined in its 1986 resolution on Community energy policy objectives that the promotion of RE was one of the EU's energy policy objectives (Council of the European Union, 1986; European Commission, 1997: 6). Yet it was not until 1993 that the EU agreed to implement ALTENER – the first EU-wide initiative to promote RE. The aim of ALTENER was to increase the share of renewables to 8 per cent (doubling the share), to treble the share of renewables in electricity generation and to ensure 5 per cent share of biofuels in transport, all by 2005 (European Commission, 1992: 24). In addition, the stated rationale for EU-level action on RE at this time was the role of the EU in coordinating national efforts, and 'ensuring their convergence towards common objectives'. Setting quantified objectives at EU level, the Commission argued, gives clear indications to consumers, producers and investors (European Commission, 1992: 21).

Beyond the harmonisation logic, climate concerns did begin to play a more significant role in the justification for RE policy development. The EU wished to show through its RE policy that 'the Community and its member states are determined to make a significant contribution to protecting the environment, and in particular reducing CO_2 emissions, by exploiting RE sources' (European Commission, 1992: 21). ALTENER was one of the few internal policies at this time that the EU could claim as climate policy. Nevertheless, ALTENER was a weak policy instrument, mostly due to a lack of allocated funds. In an effort to strengthen RE policy, the Commission followed up with a green paper on renewable energy in 1996 (European Commission, 1996) and in 1997 with a white paper (European Commission, 1997), which paved the way for policies adopted in the 2000s.

Climate policy integration into EU RE policy

Measuring CPI in the policy process and output of EU RE policy requires identifying benchmarks. In the policy process, very high levels of CPI would mean that climate policy objectives gain high priority, and even precedence over other policy objectives. We could thus expect that the most responsible actors in the policy process would be those most in favour of strong action on climate change. Thus, we could see DG Environment as the lead drafters of the policy proposals in the Commission, with the ENVI committee drafting the report for the Parliament's readings, and the environment Council formation negotiating and agreeing the Council's position. We would expect external pro-climate stakeholders (NGOs, RE industry, etc.) to have easy access to provide input to the policy process, and pro-climate arguments would receive general backing in policy circles. Additionally, the main motivation of policymakers to advance RE policy would be to achieve the long-term climate policy objectives. In this respect, the recognition of the functional interrelations between the policy areas in the process would help push for more ambitious policy measures to improve the share of RE in EU final energy consumption (given that increases in the shares of most types of RE can directly displace fossil fuel-generated energy, thus reducing GHG emissions). These expectations represent a qualitative benchmark for very high levels of CPI in the policy process of EU RE policy.

For CPI in the policy output, the distance between the policy output (or the agreed final policy objective) and expected levels of RE shares under long-term climate policy objectives could represent a clear measurement for CPI. Long-term climate policy objectives agreed in the EU are summarised in two overarching goals: first, to ensure that global temperature increases do not exceed 2° Celsius (Council of the European Union, 1996; European Commission, 2007d) and second (and flowing from the first objective) to reduce GHG emissions in the EU by between 80 and 95 per cent by 2050, compared to 1990 levels (European Council, 2009). Thus, understanding the role that RE will play in 2050 can help in understanding whether climate policy objectives are sufficiently integrated into RE policy to achieve the 2050 goal. In RE policy, very high levels of CPI would imply ambitious policy to increase the share of RE in final energy consumption in the EU. Most sources of RE can displace fossil fuels, and thus reduce GHG emissions. Therefore, it can generally be said that the more ambitious RE policy is, the better this is for achieving climate policy goals.

Many scenarios outlining the road to decarbonisation by 2050 highlight the significant role to be played by RE in the energy mix to achieve the climate goals (ECF, 2010; EREC and Greenpeace, 2010; EREC, 2010; European Commission, 2011a; Heaps, Erickson, Kartha and Kemp-Benedict, 2009), with some scenarios outlining possible pathways to a 100 per cent RE supply (overall or for the power sector) (EREC, 2010; PricewaterhouseCoopers, 2010; WWF, 2011). Other analyses include a range of solutions including CCS technologies, and nuclear energy in their assessments (ECF, 2010; European Commission, 2011a; Odenberger and Johnsson, 2010; Reichardt, Pfluger, Schleich and Marth, 2012). Depending on the assumptions regarding CCS and nuclear energy, most scenarios nevertheless imply an RE share of between 55 and 100 per cent by 2050 as required for decarbonisation. With CCS technologies still commercially risky in 2014 (DG Climate Action, 2014; Reichardt et al., 2012), and nuclear energy continuously facing public opposition for both environmental and safety reasons, I follow here the arguments that a high proportion of RE is required in the overall energy mix for 2050. Scenarios that limit or exclude both nuclear and CCS technology suggest that close to 100 per cent of our energy demands can be supplied by RE sources. Thus, taking the top ranges outlined in studies on decarbonisation to 2050, that limit (or exclude) the role of nuclear energy and CCS, suggests between 80 and 100 per cent of RE share by 2050.

Figure 3.2 shows what a linear trajectory for CPI towards 80 to 100 per cent share of RE in the EU would look like from 2000 to 2050. A linear trajectory for CPI may seem a simplistic tool for measuring the share of RE in final energy consumption to 2050, as early action requires high upfront costs. However, it can

66 *Renewable energy*

also be argued that early action is required to ensure GHG emissions peak early enough to mitigate climate change (IPCC, 2007). Thus, I use a linear trajectory to balance the effort over time. This is compared to business-as-usual (BAU) scenarios from the Commission's 2007 renewable energy roadmap and from the Commission's 2011 energy roadmap (European Commission, 2007a, 2011c). A share of RE of between 80 to 100 per cent by 2050 implies an increase by about 7 to 9 percentage points every five years from 2000. In 2000, the share of RE stood at nearly 8 per cent (European Commission, 2011c), and this increased by just one percentage point to nearly 9 per cent in 2005 (EEA, 2008: 44). The Commission's BAU scenario from 2006 includes the early 2001 RES-E Directive measures, but still only expects increases in the share of RE of about 1 percentage point every five years – hitting about 10 per cent in 2010 and just over 12 per cent in 2020 (European Commission, 2007a: 7). The second (2011) BAU scenario includes measures from the 2009 RE Directive that aim to increase the share of RE in the EU to 20 per cent by 2020 (see below). This BAU scenario outlines the achievement of the 20 per cent goal to 2020, but does not expect increases in the share of RE beyond 2020 of much more than 1 percentage point every five years – reaching just over 25 per cent by 2050 (Dupont and Oberthür, 2012; European Commission, 2011c).[1]

Based on this linear trajectory, I will be able to provide a broad assessment of the distance between the policy output of the 2001 renewable sources of electricity Directive and the 2009 RE Directive, and the trajectory for achieving very high levels of CPI.

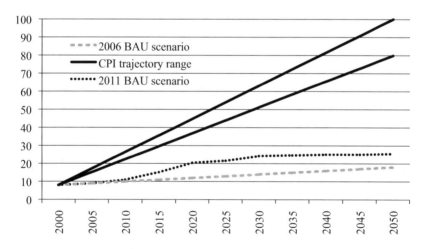

Figure 3.2 Expected very high levels of CPI from 2000–2010, compared to BAU scenarios from 2006 and 2011 (measured as a percentage share of final energy consumption)

Source: Compiled from (European Commission, 2007a, 2007b, 2011a, 2011b, 2011c), own calculations

EU RE policy in the 2000s

In the 2000s, the EU agreed on two main policy instruments for increasing the share of RE in energy consumption – the 2001 Directive on the promotion of electricity from renewable energy (2001/77/EC, called the 'RES-E Directive') and the 2009 Directive on the promotion of the use of energy from renewable sources (2009/28/EC, the 'RE Directive'). The main lines of these two policy instruments grew from debates and discussions in the 1990s.

The weakness of the ALTENER policy instrument agreed in the 1990s led to early calls for new policy development. The Commission published its white paper on 'energy for the future' in 1997, outlining several co-benefits and motivations for future policy development (European Commission, 1997; Howes, 2010: 117). These included the following points:

- RE can reduce energy dependency and increase energy security;
- The development of RE sources can promote job creation;
- RE can help the EU comply with environmental protection requirements and commitments at both EU and international-level;
- The growing energy consumption of developing countries provides (world-leading) EU renewable industries with an opportunity to expand (European Commission, 1997: 4);
- EU-wide RE policy was essential 'to avoid imbalances between members states or distortion of energy markets' (European Commission, 1997: 6–7).

The EU adopted its first Directive on promoting renewable sources of electricity in 2001. The Directive sets out an objective to achieve 22 per cent share of renewable energy sources in electricity in the EU by 2010 (revised to a 21 per cent share with the accession of new member states in 2004 and 2007).

Discussions following the 1997 white paper informed the debate leading up to the proposal and adoption of the 2001 RES-E Directive. Many of the issues of debate in these late years of the 1990s and early years of the 2000s continued to play a role in RE policy development into the future. Among the main points of contention were questions about whether targets to increase the share of renewables should be binding or not; what sources of energy can fall under the definition of 'renewable'; and whether or not support measures for RE should be harmonised at the EU level (Rowlands, 2005: 966). The Commission favoured harmonised RE certificate trading to support RE development over the (generally successful) national feed-in tariff schemes (a fixed-price payment) in place in, for example, Germany, Denmark and Spain (Boasson and Wettestad, 2013: 82). But in the face of considerable opposition, the Commission eventually put aside its insistence on introducing harmonised support schemes (Jansen and Uyterlinde, 2004: 97).

In the light of these heated debates and discussions, the publication of the proposal for a directive on RE sources in electricity was delayed from the planned date of 1998 and finally published on 10 May 2000 (European Commission,

1997: 34, 2000b). The proposal came after the EU's commitment to reduce its GHG emissions by 8 per cent by 2008–2012 compared to 1990 levels under the Kyoto Protocol (1997) had been agreed. Policy to increase the share of RE in the EU was one of the concrete responses from the EU to achieve this target (Boasson and Wettestad, 2013). At this point in time, RE policy focused on the electricity sector – the heating and cooling sector was not a part of the Commission's proposal and the transport sector was dealt with in the 2003 Biofuels Directive (2003/30/EC). The legal basis for the proposal was Article 95 of the Treaty establishing the European Community (Amsterdam Treaty, TEC), placing the RES-E Directive under the internal market competence of the EU. This reflects the debate within the Commission at the time, especially on the harmonisation of support schemes, and the focus on market integration. As the internal market for energy was developing (after the adoption of the 1996 Electricity Directive 96/92/EC and the 1998 Gas Directive 98/30/EC), the Commission feared that many different national support schemes for RE would represent barriers to fair competition and distort trade (Boasson and Wettestad, 2013; Lauber, 2001: 299; Nilsson, 2011: 115). There was much debate on what system was most effective for promoting RE, with criticism of the Commission's market-oriented approach coming from member states (citing the subsidiarity principle), environmental NGOs and RE industry (arguing that the quota system of the certificates would be ineffective). In the end, the Commission proposed to 'monitor' the national support schemes (European Commission, 2000b: 20).

The Commission's proposed definition of 'renewable energy sources' was 'renewable non-fossil sources (wind, solar, geothermal, wave, tidal, hydroelectric installations with a capacity below 10 MW, and biomass, which means products from agriculture and forestry, vegetable waste from agriculture, forestry and from the food production industry, untreated wood waste and cork waste)' (European Commission, 2000b: 19). The Commission did not propose to make the 12 per cent target legally binding, however, and instead requested five-yearly updates from member states on their plans, with the Commission scrutinising member state progress on an annual basis. In the event that the Commission found a lack of progress in member states, it should be able to present proposals for imposing mandatory targets (pp. 19–20). The proposed indicative targets for the share of RE in electricity production per member state were outlined in the annex to the proposal. The biggest increase was expected from Denmark (with a target of 29 per cent of RES-E in 2010 – an increase from an 8.7 per cent share in 1997). Discussions took place in advance of the publication of the proposal with member state agencies, industry representatives, professional associations and NGOs, and among the Commission services (European Commission, 2000b: 13). In all of these consultations, the focus was on support schemes. Consultation also took place with the electricity supply industry and associations representing the RE industry on issues of administrative barriers and grid reinforcement requirements for integrating more RE into the electricity system.

The Parliament's energy committee (ITRE) was in charge of drafting a first reading response to the RES-E proposal. MEP Mechtild Rothe of the European

Socialists political group was the rapporteur. MEP Hans Kronberger (no political group) provided an opinion on behalf of the ENVI committee. The first reading report was adopted by the Parliament on 16 November 2000, in which the Parliament suggested 68 amendments to the Commission's proposal (European Parliament, 2000b). The main substantive changes included the Parliament's disagreement with the Commission's calculations. The Parliament argued that the 12 per cent overall share by 2010 translated into a 23.5 per cent share of RE in electricity generation (as also mentioned in the 1997 white paper; European Commission, 1997), and requested a change in indicative targets in this respect. Additionally, the Parliament called for the national targets to be mandatory. Finally, the Parliament supported national-level support schemes. The rapporteur explained that there was a positive experience with the support schemes, and that they were necessary because of historical support to conventional and nuclear energy production that led to unfair competition for the new RE producers (European Parliament, 2000b). Generally, the Parliament aimed to strengthen the proposal in favour of renewable electricity and ensure trade and competition concerns did not hamper RE development.

The Commission responded with an amended proposal in December 2000 (European Commission, 2000a). It accepted a number of the Parliament's proposed amendments, but it did not accept Parliament's calculation of 23.5 per cent share in electricity (p. 5). Nor did the Commission take on board amendments making the national targets mandatory, but kept the phrase that the Directive would '… require all member states to set national targets …' (p. 6). The Commission otherwise accepted much of the wording of the Parliament about evaluating national support schemes.

The Council's common position was the next step in the legislative proposal, and it was published on 23 March 2001 (Council of the European Union, 2001). The Council placed the proposal under the environmental chapter of the Treaty (Article 175.1, TEC), rather than under the internal market, as proposed by the Commission. The Council clarified that the targets should be non-binding, stating 'all member states should be required to set national indicative targets for the consumption of electricity produced from renewable sources' (p. 3). When drawing up their own 'indicative' targets, member states had to 'take account of' the national targets drawn up per member state by the Commission in the annex to the proposal (p. 9). Additionally, the Council removed the limit on hydropower in the definition of renewable energy sources (from a maximum of 10 megawatts), so that even large-scale hydropower could be considered a renewable energy source.

The proposal moved into second reading in the Parliament. The ITRE committee tabled the second reading recommendation in the plenary on 20 June 2001 (European Parliament, 2001). Media predicted a tough battle between Parliament and Council in the second reading, with MEP and rapporteur Mechtild Rothe warning that she would stick to the proposal for legally binding targets (ENDS Europe, 2001b). Informal negotiations between the then Swedish Presidency and the rapporteur resulted in a deal aimed to avoid entering the conciliation procedure (or the third reading) (ENDS Europe, 2001a). The negotiated deal meant that the non-binding targets remained in place in exchange for a commitment to introduce binding targets in future if the indicative approach fails. As amended by the Parliament in its second reading, recital seven of the final Directive reads: 'if necessary for the achievement of the targets, the Commission should submit proposals to the European Parliament and the Council which may include mandatory targets'. Agreement was also finally reached between Parliament and Council to allow energy produced from waste incineration and hydropower (without any upper capacity limit) to be defined as renewable energy (Art. 2). The final act was signed by Council and Parliament on 27 September 2001.

The Commission reported regularly on the progress in achieving the targets set out in the 2001 RES-E Directive in the years that followed. In its 2006 review, the Commission highlighted less-than-perfect implementation and paved the way for future policy measures. By this time, it had already begun infringement proceedings against six member states (Austria, Cyprus, Greece, Ireland, Italy and Latvia) for reasons of incomplete transposition of the Directive into national law; lack of commitment on the targets; lack of implementation of the guarantees of origin certificate system; lack of transparency in administrative procedures to issue licenses for new renewable electricity plants; and lack of transparency regarding access to grids and regarding rules on grid investment (European Commission, 2006a: 18). As a result of the poor performance in improving the penetration of RE sources of electricity, the Commission announced it would publish a roadmap for RE and propose a new legal framework (p. 19).

The 2009 Directive on increasing the share of renewable energy in the EU's final energy consumption outlined an objective of achieving a 20 per cent share of RE by 2020 in the EU, with specified national targets per member state. These targets are legally binding on member states.

The promised renewable energy roadmap was published in 2007. It aimed to establish a long-term vision for RE in the EU (European Commission, 2007b). It highlighted the unlikelihood of meeting the 2010 target, which constituted a 'policy failure and a result of the inability or the unwillingness to back political declarations by political and economic incentives' (p. 8). Most importantly, the roadmap contained a proposal for a legally binding target of 20 per cent share of RE in the EU's final energy consumption by 2020 – a target that remained throughout the later negotiation process and became part of the final text of the RE Directive 2009/28/EC (p. 3). This target came in response to scenarios and assessments in light of the European Council's call for a target of 15 per cent share of RE by 2015, and the European Parliament's call for a 25 per cent target by 2020 (European Council, 2006: 15; European Parliament, 2006). The roadmap also outlined a departure from earlier legislation by highlighting the need for the future policy framework to cover RE sources in electricity, transport and the heating and cooling sector under one policy instrument (with the RE target then being measured as a share of final energy consumption).

The Commission put forward its proposal for a Directive on the promotion of the use of energy from renewable sources in January 2008 (European Commission, 2008). It was proposed as part of the 'climate and energy package' that also included a proposal for revising the emissions trading system (ETS), a proposal for a decision on reducing GHG emissions in sectors not covered by the ETS, and a proposal for a Directive to support CCS technology (Oberthür and Pallemaerts, 2010a, 2010b). The proposal established a binding 20 per cent RE target for 2020, and a second target to increase the share of biofuels and renewables in transport to 10 per cent by 2020 (European Commission, 2008: 8). It was proposed on the dual legal basis of the environmental chapter, Article 175.1 TEC, and Article 95 TEC on the internal market. The Commission justified this dual legal basis as the proposed articles on biofuels and bioliquids prevent member states from adopting measures that would block trade in biofuels. The rest of the proposed Directive is considered to fall under the objectives of the environment chapter to 'preserve, protect and improve the quality of the environment, protect human health and make prudent and rational use of natural resources' (p. 8).

The consultations leading up to the proposal included public consultations on the 2007 RE roadmap, on the energy green paper (European Commission, 2006b) and on the strategic energy review (European Commission, 2007c) between March and September 2006. In addition, public consultations took place in 2007 with 'member states, citizens, stakeholder groups, civil society organisations, NGOs and consumer organisations' (European Commission, 2008: 5). The major issues touched upon in these public consultations included a review of the promotion of biofuels in new RE legislation; the promotion of heating and cooling from RE; and administrative barriers to the development of RE. The Commission reported general support for stronger policy and long-term goals in RE policy, and also support for sustainability criteria for promoting biofuels.

The process leading to the adoption of the 2009 RE Directive took place in one reading. The negotiations among the institutions began informally early after the publication of the Commission's proposal. In the Parliament, MEP Claude Turmes (Greens) was the rapporteur for the ITRE committee on the dossier, but the ENVI committee had a special role as an associated committee for the first reading report (meaning it was jointly responsible with the ITRE committee, and did not simply provide an opinion). Anders Wijkman of the EPP group was the rapporteur for the ENVI committee. MEPs highlighted three main issues in the proposal that they would like to see strengthened: the interim targets for member states to meet the 2020 target (including sanctions if such interim targets were not met); sustainability criteria for biofuels in transport, if that target was to remain; and priority access for renewables to the electricity grid (ENDS Europe, 2008j). A fourth issue, that was a sticking point also for Council, was on the renewables trading certificates (ENDS Europe, 2008a, 2008f). While the rapporteur considered the proposed Guarantees of Origin (GO) certificate trading too legally ambiguous, member states (also supported by the Parliament) called instead for more flexibility on renewables trading (ENDS Europe, 2008a; Nilsson, Nilsson and Ericsson, 2009).

Throughout 2008, negotiations between Parliament and Council continued in trialogues. The proposal was discussed in the energy Council formation on 28 February, 6 and 9 June and 8 December 2008. Although the environment Council discussed the climate and energy package on 3 March, 5 June, 20 October and 4 December 2008, it focused on the emissions trading system proposal, the effort-sharing decision and the CCS proposal. These were the three legislative proposals considered within the 'competence' of environment ministers (Council of the European Union, 2008c: 9). The main points of contention for member states in the negotiations on the RE Directive included the strength of the national targets (and the consequences of not meeting the indicative trajectory), the sustainability criteria for the biofuels target and the system of GO trading (Council of the European Union, 2008a, 2008b; ENDS Europe, 2008a).

In September 2008, the ITRE committee insisted in the informal negotiations on mandatory interim targets for member states, with penalties for missing these targets (ENDS Europe, 2008g). The committee was ready to accept Council's proposal for more flexibility in renewables trading and the 10 per cent target for biofuels and renewables in transport. However, the committee included certain criteria for biofuels before they could be considered under the target (that at least two-fifths of the overall share should be from second-generation biofuels or electric vehicles powered by RE). A further review clause on the biofuels provisions for 2014 was included. The rapporteur praised the inclusion of this review clause as an opportunity to strengthen the Directive later (ENDS Europe, 2008h).

By the end of October 2008, member states had reached an agreement within the Council, and negotiations on the remaining sticking points with Parliament continued. The major remaining issue at this point was the sustainability criteria for biofuels (Council of the European Union, 2008b; ENDS Europe, 2008d). Additionally, a review of all the targets in the Directive for 2014 was pushed for by Council – some member states saw this review clause as an opportunity to reconsider the strength of the targets overall (Council of the European Union, 2008e). Compromise between the Parliament and Council was reached by early December. Interim targets to 2020 remained indicative and no automatic financial sanctions were included for member states that miss their binding 2020 target. Agreement was reached that measures on the indirect land use changes due to biofuel production would be adopted later and the Commission was granted oversight over member state renewable action plans to be submitted after the Directive was adopted. Despite Italian desires to the contrary, the 2014 review clause clarifies that no changes can be made to the RE targets as a result of the review (ENDS Europe, 2008c, 2008i).

With the RE Directive forming part of the climate and energy package, it was negotiated alongside the revised ETS Directive (2009/29/EC), the Directive on CCS (2009/31/EC) and the effort-sharing Decision for emissions not covered by the ETS (Decision 406/2009/EC). On 12 December, in an unusual move under the ordinary legislative procedure, the European Council announced agreement on the entire package of legislative measures (ENDS Europe, 2008e; European Council, 2008a: 8–9). This announcement came before the Parliament adopted its first reading agreement, on 17 December 2008 – although the positions were already aligned, the Parliament had not yet officially adopted its position in plenary. The Commission accepted the text and the Directive was signed into law on 24 April 2009 as Directive 2009/28/EC.

CPI into the policy process of the 2001 RES-E Directive (2001/77/EC)

Having discussed the general outputs and processes leading to the adoption of the 2001 RES-E Directive, I turn now to measuring the level of CPI in the policy process. As discussed above, and in Chapter 2, I examine three elements of the policy process to measure the level of CPI: the involvement of internal and external pro-climate stakeholders in the policy process, and the recognition of functional interrelations with long-term climate policy objectives. Together, these indicators will provide a qualitative assessment of the level of CPI in the process, in line with Table 2.1.

Internal pro-climate stakeholders

In the Commission, DG Transport and Energy ('DG Energy') drafted the proposal for the RES-E Directive. Much of the discussion within the Commission in advance of the (delayed) publication of the proposed Directive centred around the idea of harmonising support schemes (ENDS Europe, 1999b; Jansen and Uyterlinde, 2004). In the end, with member states, RE industry associations, environmental NGOs and MEPs all arguing against harmonising support schemes, the Commission accepted to 'monitor' national support schemes (Boasson and Wettestad, 2013; European Commission, 2000b). There is no clear evidence of the pro-active involvement of DG Environment in the development of the Commission's proposal, but they were certainly involved in the long discussions before the proposal was published.

In some respects, DG Energy itself can be said to have acted as a pro-climate stakeholder in the policy process. There may not have been any further need for the involvement of DG Environment, as the aim of the Commission was to advance RE policy in order to achieve climate objectives, at least in the short-term (under the Kyoto Protocol). In the proposal, the Commission states: 'In view of the substantial contribution RES can make to the implementation of the Community's commitments to reduce greenhouse gases, its expansion in the EU constitutes an essential part of the package of measures needed to comply with the Kyoto Protocol' (European Commission, 2000b: 3). With such a pro-climate stance, even without DG Environment taking the lead, we can assume DG Energy played a pro-climate role internally in the Commission. As suggested in Table 2.1, this involvement of a pro-climate stakeholder internally in the Commission implies a *medium* level of CPI in the policy process.

In the Parliament, although the ITRE committee drafted the report, the ENVI committee provided an opinion for the rapporteur. In this case, some of the ENVI suggestions were taken on board, and overall, it can be said that Parliament was ambitious in pushing for a policy measure that would make a difference, at least in the first reading. The Parliament pushed for mandatory national targets and highlighted the environmental and climate policy objectives of strong RE policy. The rapporteur's first reading report explained the many advantages of pushing for ambitious RE policy as follows:

> Renewable energies are an integral feature of an effective strategy to protect the climate; they help achieve the Kyoto objectives; they do not waste resources; they reduce emissions of harmful substances into the air; they make it possible to develop a decentralised structure of energy supplies, together with the possibility of sustainable regional development and new employment prospects; they create security of supplies; and, in an international context, they help developing countries solve a variety of problems.
> (European Parliament, 2000b: 38)

Here, it is clear that the climate and environmental objectives that can be achieved through increasing RE are of paramount importance for the ITRE committee, with co-benefits for job creation and energy security coming later in the list of benefits. It could thus be said that the rapporteur in this case (MEP Mechtild Rothe) was a pro-climate stakeholder herself, and that the ITRE committee was committed to advancing climate policy through RE policy. The ENVI committee did also have the opportunity to provide an opinion and influence the final ITRE committee report. ENVI's amendments highlighted the role of the RES-E Directive in meeting GHG emission reduction targets under the Kyoto Protocol, which was accepted in the final legislative act. Other proposals, taken on board also by the ITRE committee (such as ensuring binding national targets) did not survive the policy negotiations. In this case, at least for the first reading, the ENVI committee provided several points that were taken on board by the ITRE committee in the drafting of the first reading report. However, negotiations on the second reading agreement took place informally between the rapporteur and the Presidency (ENDS Europe, 2001a). The ENVI committee had no role in these informal negotiations. Even though MEP Mechtild Rothe aimed to keep her strong stance throughout the second reading (ENDS Europe, 2001b), she finally relented and accepted many member state demands for the sake of agreement. In accordance with Table 2.1, the involvement of internal pro-climate stakeholders in the Parliament is *medium*. This is because, although ENVI provided an opinion in the first reading report, and several of ENVI's proposed amendments were included, ENVI did not participate in the informal negotiations that followed in second reading. While the rapporteur originally claimed she would stick to the objective of making the targets mandatory, she finally compromised with member states in exchange for promises that proposals for mandatory targets could be made in future if the indicative approach failed.

In the Council, negotiations took place within the energy Council formation. There is no apparent interaction with environment ministers (although it is true that negotiations in Council are often behind closed doors). Even though the Council and the European Council supported the objective to increase the share of RE in the EU to 12 per cent by 2010, as outlined in the 1997 white paper (see above), member states were not prepared to agree to mandatory targets. This, and the issue of the definition of sources of renewable energy, seemed to be the main sticking points in the negotiations with Parliament (ENDS Europe, 2000d). The French Presidency hoped to reach rapid agreement (ENDS Europe, 2000b), but these issues delayed agreement until 2001. In early 2000, some member states already raised questions about their shares of the indicative target, calling for them to be reduced. Italy pushed for waste incineration to be included as part of the definition of renewably-sourced electricity – a proposal that Parliament did not accept in the early stages of the negotiations (ENDS Europe, 2000d). Finally, the Council pushed for waste incineration as part of the definition and supported indicative targets. Portugal, Finland and the Netherlands negotiated for lower indicative targets to 2010 as the proposed targets were considered too ambitious (Council of the European Union, 2001; ENDS Europe, 2000a). Unlike in the Parliament and Commission, the absence of pro-climate stakeholders in the policy process that could pressure for high political commitment to strong RE policy may have lowered the level of CPI overall in the policy process in the Council. In this case, (and in line with Table 2.1) there is *very low* evidence of strong involvement of internal pro-climate stakeholders in the policy process.

Taking an aggregate of the *medium* levels of involvement of pro-climate stakeholders in the Commission and Parliament with the *very low* levels in the Council, the overall level of involvement of internal pro-climate stakeholders in the process leading to the agreement on the RES-E Directive is *low to medium*.

External pro-climate stakeholders

The Commission's proposal for the RES-E Directive was published in 2000, before requirements for an impact assessment and open public consultations existed. These became a regular part of the Commission's preparations for policy proposals after 2002, when better regulation and minimum standards for consultation were adopted (European Commission, 2002). No impact assessment procedure or open public consultations leading to the publication of the Commission's proposal were carried out. Nevertheless, there was much discussion among EU institutions, member states and other stakeholders in the aftermath of the publication of the 1997 white paper. The Commission reports that industries, professional associations and NGOs were involved, along with Commission officials and member state representatives, in the discussions (European Commission, 2000b: 13). RE industry representatives were consulted, especially with regard to administrative procedures and grid issues.

Without formal consultation procedures in place, it is difficult to assess the extent to which external pro-climate stakeholders had access to the Commission. One interviewee suggested that in the late 1990s and early 2000s, environmental NGOs, in particular, were less a part of the policy process in general than in the mid- to late-2000s (interview 7). The Commission was not always the first lobbying target for environmental NGOs. The RE industry at this time was focused on ensuring that national support schemes remained in place, and rather worked with the Parliament and through their national governments, than with the Commission (Boasson and Wettestad, 2013; Hildingsson et al., 2010). The German RE industry was particularly strong on pushing the German government not to accept Commission proposals to harmonise support schemes (ENDS Europe, 1999b). A coalition of environmental groups did try to influence the Commission's proposal by calling jointly for a strong RE law and pushing EU member states to agree to a 16 per cent RE share target for 2010 (ENDS Europe, 1999c). The coalition, including Greenpeace and WWF, targeted member states instead of the Commission. Later, the NGOs WWF and Climate Action Network criticised the Commission's low ambition in its proposals (ENDS Europe, 1999a). It seems that the informal lobbying of environmental NGOs in the advance publication of the proposal did little (if anything) to encourage the Commission to propose more ambitious policy.

Thus, the involvement of external pro-climate stakeholders in the policy process with the Commission is *low*. There were no official procedures in place to allow external stakeholders access to the Commission, and consultation seems to have taken place on an ad hoc basis. The involvement of environmental NGOs is unclear, and the RE industry preferred to lobby their own governments first.

When it comes to the Parliament, environmental NGOs generally agree that they have easy access, and better chances to make their voices heard (interviews 5, 6, 7, 8 and 11). In the early 2000s, environmental NGOs backed up the Parliament's first reading stance on the RES-E proposal: namely, to push for mandatory targets and to ensure that national support schemes continued. At this time, the RE industry was becoming increasingly organised at the EU level in the form of the European Renewable Energy Council, EREC (Boasson and Wettestad, 2013). The Parliament was generally an ally of environmental NGOs and the RE industry, who were reportedly happy with the result of the first reading (ENDS Europe, 2000c). The same external pro-climate stakeholders were, however, generally dissatisfied with the final result of the policy negotiations (Nilsson, 2011). Nevertheless, although Parliamentary actors were overall in favour of pushing for more ambitious policy on RE, there were no official procedures for consultation. The involvement of external pro-climate stakeholders was on an informal basis. While interviewees considered their access to the Parliament as sufficient (interviews 1, 5, 6, 7, 8 and 11), their access to policymaking processes became more limited as the process continued. Informal negotiations in the second reading effectively excluded external pro-climate stakeholders. They could not access the policymaking process and neither could they follow the developments in the process, as negotiations took place behind closed doors (interview 6).

According to Table 2.1, this situation amounts to a *low to medium* level of involvement of external pro-climate stakeholders in the policy process with the Parliament. There were no official procedures in place, but all interviewees reported easy access to the Parliament. However, this ease of access was only in the early stages of the policy process, as informal negotiations between Council and Parliament in the second reading effectively excluded possibilities for external stakeholder involvement.

In the Council, there were few opportunities for external stakeholders to get involved. Many negotiations in the Council take place behind closed doors, and are not transparent (interview 6). Environmental NGOs were particularly limited in this case in their ability to access the Council, which often requires lobbying governments at the member state level. Several NGO interviewees commented that their own internal resources did not allow them to follow such a strategy (interviews 5, 6 and 7). The RE industry, however, lobbied at both member state and EU level. RE industries were organised at EU level under EREC, but they did not benefit from any special access to the Council. Their push for binding RE targets, for example, did not survive in the Council (Council of the European Union, 2001).

The involvement of external pro-climate stakeholders in the process with the Council can only be considered *low*. While in the early stages, RE industries in some member states could access their national governments and push policy at that level, in later stages of the policy negotiations, there was no access to the Council for external environmental NGOs or RE industry representatives. These stakeholders tried to remain visible throughout the process, but could not access a closed Council.

In summary, the involvement of external pro-climate stakeholders in the policy process leading to the adoption of the 2001 RES-E Directive is generally *low*, although informal access to the Parliament scored *low to medium*.

Recognition of functional interrelations

From the very beginning of the development of RE policy in the EU, policymakers highlighted the benefits of RE promotion for climate policy. The Commission stated that one of the main objectives of the RES-E Directive was to meet 'the obligation to reduce the emission of greenhouse gases accepted by the EU at Kyoto' (European Commission, 2000b: 2). The Parliament also highlighted, especially, the climate and environmental objectives of increasing the share of RE in the EU, and aimed to strengthen the Commission's proposal (European Parliament, 2000a). The Council and European Council had originally supported the Commission's 1997 white paper on renewable energy, calling for a doubling of the share of RE in the EU to 12 per cent by 2010 as a response to climate commitments. Yet in the negotiations on the RES-E Directive, member states watered-down the proposal to ensure targets were 'indicative' and that waste incineration would be defined as a renewable source of energy (Council of the European Union, 2001).

The synergetic functional interrelations between RE policy objectives (to increase the share of RE in EU energy consumption) and climate policy objectives (to reduce GHG emissions) were certainly recognised by policymakers in the early stages of the policy process especially. The recognition was rather limited to short-term climate policy objectives (such as the commitments under the Kyoto Protocol for 2008–2012), and there is little mention of long-term climate policy objectives to limit global temperature increase to 2° Celsius. The final Directive does not provide for mandatory targets, and neither can the output (12 per cent RE share by 2010) be considered ambitious from a long-term perspective (see below). Policy negotiations in the later phases pitted the Parliament against the member states. In some respects, the functional interrelations remained a stronger part of the negotiation strategy of the Parliament, yet the rapporteur nevertheless compromised with the Council in informal negotiations in the second reading. Perhaps for the sake of getting agreement, the long-term functional interrelations (which, if recognised, should help push for ambitious RE policy), were rather left aside in the later stages of the policy negotiations.

As outlined in Table 2.1, the recognition of functional interrelations between long-term climate policy objectives and RE policy objectives in the policy process leading to the RES-E Directive is situated close to the *medium* level (functional interrelations recognised and considered by policymakers in the process, but this does not motivate sufficiently strong policy action). The co-benefits of the policy objectives were highlighted from the outset of the policy proposal, yet related more to short- or medium-term climate policy objectives than long-term objectives. The recognition of these functional interrelations turned out to be insufficient to push policy in a more ambitious direction.

Taken together, the *low to medium* levels of involvement of internal pro-climate stakeholders, the *low* levels of involvement of external pro-climate policy stakeholders, and the *medium* recognition of the functional interrelations in the policy process suggests an overall level of CPI in the policy process in the 2001 RES-E Directive of *low to medium*.

CPI in the policy output of the 2001 RES-E Directive

As shown in Figure 3.2, very high levels of CPI for RE policy would see an increase in the share of RE in EU final energy consumption by between 7 and 9 percentage points every five years from 2000 to meet long-term climate objectives by 2050. We can use data recorded on the shares of RE in the EU since 1990 to make suggestions of what BAU between 1990 and 2010 might have looked like without the RES-E Directive in place. Plotted against the expected BAU scenario, the 12 per cent target as the policy output for the RES-E Directive to 2010 shows the distance covered by this policy instrument towards closing the gap between BAU and very high levels of CPI.

According to the EEA, the share of RE in gross inland energy consumption in the EU27 stood at 4.2 per cent in 1990, rose to 5 per cent in 1995 and to 5.6 per cent in 2000 (see EEA online database: eea.europa.eu). This implies that the share of RE in the EU was on the increase before the 2001 RES-E Directive was agreed, but only by about 0.7 percentage points every five years. Assuming this trend would have continued as BAU between 2000 and 2010, there would have been an increase in RE share in gross inland energy consumption to 6.3 per cent in 2005, and to 7 per cent in 2010 (European Commission, 1996: 15). This is clearly below the 2001 RES-E Directive's target. With RE share of 5.6 per cent in 2000, meeting the 80 to 100 per cent CPI targets suggests a required percentage share of between 21 and 25 per cent share in RE by 2010 (see Figures 3.2 and 3.3). Thus, although the 2001 RES-E Directive's policy output succeeds in closing the gap between BAU and CPI trajectories by 5 percentage points, this is still far from the 2050 trajectory. In other words, the policy output of the 2001 RES-E Directive is between 28 and 36 per cent in line with long-term climate policy objectives to decarbonise by 2050. According to Table 2.2, this implies *low* levels of CPI for the policy output in the 2001 RES-E Directive (see Figure 3.3).

As the targets of the 2001 RES-E Directive were indicative targets for 2010, it is now also possible to examine the actual implementation of the Directive. According to Commission data from 2011, the EU achieved a 9 per cent share of RE in 2010, so the 12 per cent target was not met (European Commission, 2011c). In this case, the actual implementation of the 2001 RES-E Directive increased the share of RE by just 2 percentage points compared to BAU, implying the RES-E Directive closed the gap between BAU and complete CPI by just 11–14 per cent (*very low* levels of CPI, see Table 2.2).

These *very low* levels of CPI in the policy output of the 2001 RES-E Directive could nonetheless also be considered too harsh a measurement. It would be useful to consider the levels of CPI in the policy output in terms of capacity and

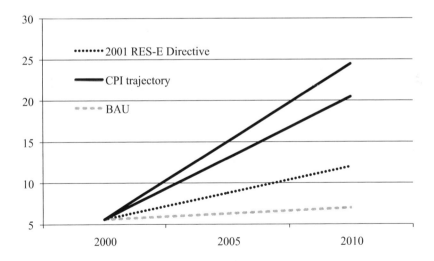

Figure 3.3 Policy output of the 2001 RES-E Directive compared to BAU and to very high levels of CPI (measured as a percentage share of gross inland energy consumption)

Source: EEA database, www.eea.europa.eu; own calculations

potential for achieving a speedy increase in the share of RE at the time. In its 1996 green paper on the future of energy, the Commission outlined one scenario that suggested about 13 per cent share of RE in gross inland energy consumption would be possible with strong policies by 2010. Considering that 9 per cent was achieved without mandatory targets (without what could be called strong policy), a 13 per cent share seems rather low. Nevertheless, it is not unrealistic to argue that RE technology was still costly in the early 2000s. Making an assumption that a steeper climb to the 2050 targets would begin after 2010, as costs for RE fall, would provide a different scenario. If the 12 per cent share was achieved in 2010, it would imply between about 9 and 11 percentage points increases every five years from 2010 to meet the 2050 target. This may be considered quite a jump from the 1 to 2 percentage point increases every five years from 1990 to 2010. Taking instead a target, as argued for by the European Parliament of 15 per cent RE share in 2010 would call for between 8 and nearly 11 percentage point increases per five years beyond 2010. A midway point between the very high CPI figures of 21 and 25 per cent RE share by 2010, and the actual 12 per cent target would suggest between about 17 and 19 per cent RE share in 2010 (leaving between 8 to 10 percentage point increases required per 5 years from 2010 to 2050).

Figure 3.4 outlines these midway ranges for CPI to 2010 against the policy output of the 2001 RES-E target and against the BAU scenario. Taking this more tempered benchmark into account, the policy output of the 2001 RES-E

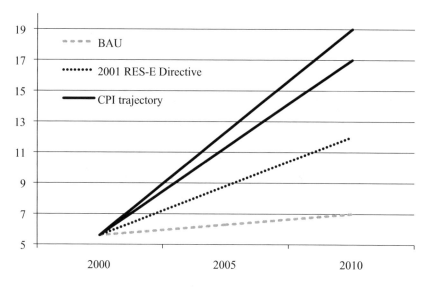

Figure 3.4 CPI in the policy output of the 2001 RES-E Directive, assuming lower advances in RE share between 2000 and 2010 than between 2010 and 2050

Sources: Compiled from (European Commission, 2007b, 2011c), own calculations

Directive closes the gap between BAU and very high CPI levels by between 42 and 50 per cent. Such a level of CPI in the policy output is *medium*.

However, considering the expected implementation deficit of many pieces of EU legislation, especially when the targets are non-binding (Haverland and Romeijn, 2007; Lampinen and Uusikylä, 1998), and considering the empirical advantage available for the RES-E Directive to assess *ex ante* the performance, even a generous level of *medium* is misleading. The final 9 per cent share of RE in the EU in 2010 closes the gap between BAU (7 per cent share in 2010) and a tempered CPI level (between 17 and 19 per cent share in 2010) by between only 20 and 25 per cent. This level, according to Table 2.2 indicates *low* levels of CPI in the policy output of the 2001 RES-E Directive.

Overall, we can say that the level of CPI in the policy output of the 2001 RES-E Directive is *low*, considering the non-binding nature of the (already insufficient) target, the recognised implementation deficit for EU legislation, and considering the final implementation outcome of the Directive.

CPI into the policy process of the 2009 RE Directive (2009/28/EC)

In this section, I discuss the levels of CPI in the policy process of the 2009 RE Directive, thereby addressing whether or not higher levels of CPI can be found in the later policy developments.

Internal pro-climate stakeholders

As the RE Directive was negotiated under the ordinary legislative procedure, consultation procedures were in place (this means internally within the EU institutions and with external stakeholders through open, public consultations). In the Commission, DG Energy drafted the proposal on the RE Directive. Usual inter-service consultation procedures applied allowing space for DG Environment (and all other DGs) to lend its voice to the drafting. Additionally, the RE Directive was proposed as one part of the climate and energy package in early 2008. Other measures in this package (such as the proposal for a Directive on CCS, and proposal for a revised ETS Directive) were under the responsibility of DG Environment. There was enhanced cooperation between DGs Energy and Environment on the preparation of these draft proposals to publish a coherent package of policy measures in early 2008 (Boasson and Wettestad, 2013: 45). As such, the level of involvement of internal pro-climate stakeholders in the Commission for the 2009 RE Directive is (according to Table 2.1) *medium to high*. This is because DGs Energy and Environment worked closely together in the drafting of the proposals in the climate and energy package, but DG Energy was solely responsible for the RE Directive proposal. However, DG Energy can also be considered to have integrated the idea of promoting RE for the purpose of combating climate change into its own workings. The Commission presented RE policy as a part of the EU's response to climate change.

In the Parliament, the ITRE committee was responsible for drafting the first reading report on the proposal, but the ENVI committee was associated with the ITRE committee for the drafting procedure. Although ITRE was still the leading committee, the associated committee status for ENVI implies high collaboration between the committees on this particular dossier. The rapporteurs of the ITRE committee (MEP Claude Turmes) and the ENVI committee (MEP Anders Wijkman) worked closely together under the associated committee procedure to finalise a first reading report. Bolstering the internal pro-climate stakeholder involvement in the Parliament further, the rapporteur on this dossier was a member of the Greens political group. MEP Claude Turmes pushed amendments that strengthened the sustainability criteria for meeting the 10 per cent target in transport, and reached a compromise solution with the Council on the review of the Directive scheduled in 2014, so that it would not involve a review to lower the ambition of the targets. With the ENVI committee holding the associated committee status, and with the ITRE committee's rapporteur being a member of the Greens political group, the involvement of internal pro-climate stakeholders in the Parliament is *high* (see Table 2.1).

In the Council, the energy and environment Councils were necessarily linked, given the fact that the RE Directive was proposed as part of the climate and energy package. The package was discussed in both the energy and environment Council meetings throughout 2008, but there was a clear division of labour in the Council, with the energy Council formation being responsible for the RE Directive. It was the energy Council under the French Presidency in the second half of 2008 that negotiated informally with Parliament (Boasson and Wettestad, 2013; Council of the European Union, 2008d; ENDS Europe, 2008k). Although the overall progress of the climate and energy package was reported to both the energy and environment Council formations, the substantive discussions on the RE Directive remained within the remit of the energy Council. In this case, following the distinction outlined in Table 2.1, the involvement of internal pro-climate stakeholders is closer to *low*. Nevertheless, it can still be argued that the energy Council demonstrated some level of CPI in its negotiations on the RE proposal, and that it had internalised some amount of CPI without needing constant involvement from the environment Council. This can especially be seen through the direct involvement of heads of state and government (through the European Council) in the negotiations, which may have pushed energy ministers to agree to the RE Directive earlier. The level of CPI in the Council could thus be regarded as standing closer to *medium* levels than to *low* levels.

As an aggregate, however, the involvement of pro-climate stakeholders in the policy process leading to the adoption of the RE Directive is *medium to high*. The *medium* levels of involvement of the environment formation in the Council on the RE Directive lowers the *medium to high* and *high* levels of involvement in the Commission and the Parliament.

External pro-climate stakeholders

As discussed above, in the lead up to the publication of the proposal for the RE Directive, the Commission had several open and public online consultations related to specific issues around RE policy. NGOs and members of the RE industry were involved and present in these public consultations (interview 5). Procedures were in place to allow them to provide input in the early stages of the drafting of the proposal. In each of the public consultations mentioned above, the proportion of environmental NGOs was generally small, due to large numbers of responses from many interested stakeholders (including member states, citizens, other industries, etc.). For example, the consultation on the Commission's energy green paper in advance of the second strategic energy review attracted 164 written comments, of which 22 were from all types of NGOs (European Commission, 2006c).

In general, pro-climate external stakeholders were most concerned with the biofuels criteria, and with ensuring interim mandatory targets for renewables. There was thus some disappointment that the Commission's proposal did not elaborate on either of these aspects, and that these points were not strengthened in the final output (ENDS Europe, 2008b; ENDS Europe, 2008k). However, the procedures for consultation allowed space for external pro-climate stakeholders to express their opinions. NGOs felt they could have access to the Commission if they wished it. As one interviewee mentioned, she was never refused a meeting with a Commission official to discuss the RE proposals between 2006 and 2008 (interview 7). The fact that the RE proposal was part of the overall climate and energy package meant that the general climate issue was often discussed at stakeholder meetings, and not necessarily with a focus on RE solely. Other RE industrial actors also agreed that, at this time, there were no barriers to access to the Commission. The Commission did not necessarily seek out input of the pro-climate stakeholders, but they were open to any requests for meetings (interviews 5 and 7).

Therefore, the involvement of external pro-climate stakeholders in the policy process with the Commission leading up to the publication of the RE proposal is *high*. Procedures were in place so that environmental NGOs and the RE industry could provide their opinions to the policy process. There was some limited opposition (from major GHG-emitting industries, reticent member states, for example) to pro-climate stakeholder arguments. In addition, access to members of the Commission on this file was considered relatively easy.

In the Parliament, most interviewees agree that access for external stakeholders was easy (interviews 5, 6, 7, 8, 25 and 26), although this ease of access was linked to personal relationships with MEPs and/or their assistants. In the case of the negotiations around the RE Directive, with Green MEP Claude Turmes acting as rapporteur on this dossier, the Parliament was particularly open to input from pro-climate external stakeholders (Boasson and Wettestad, 2013; interview 7). The rapporteur also created a group of stakeholders, including NGOs and representatives of the RE industry, to facilitate the drafting of the committee's report, with a special focus on the issue of renewables trading (Boasson and Wettestad, 2013: 91). The Parliament was therefore an open institution for external pro-climate stakeholders, with the rapporteur (as a green MEP) being particularly open to opinions of environmental NGOs and the RE industry. Although the negotiations in the policy process (especially in the later stages) involved much informal negotiating behind closed doors, contact between external pro-climate stakeholders and Parliamentary actors was nonetheless considerable (interview 7). In sum, and according to the indicators described in Table 2.1, the level of involvement of external pro-climate stakeholders in the Parliament is *high*, with easy access to the policy process in the Parliament.

In the Council, it is generally noted that access for external pro-climate stakeholders is difficult (Hauser, 2011; interviews 5, 6, 7, 8 and 11). In the case of the RE Directive, however, there were certain member states that were more interested in strong and ambitious policy than others. Environmental NGOs described that they made efforts to target these member states, even in the years before the proposal was published (interview 7). With the Council, some of the conventional power industries had strong lobbies at national level to support a system of tradable certificates that was seen as disadvantageous for RE development by environmental NGOs and the RE industry (Boasson and Wettestad, 2013; interview 7). Pro-climate stakeholders in this case had some success in targeting specific member states that shared their opinions. Nevertheless, except on the national level and by targeting specific national governments, there was little to no access to the Council for external pro-climate stakeholders. Additionally, with the proposal being agreed in first reading, the process was rapid and much of the negotiating took place informally and behind closed doors. There was no room for external pro-climate stakeholders to take part in such informal negotiations. The level of involvement of external pro-climate stakeholders in the policy process in the Council is thus *low*.

Taken together, the *high* levels of involvement of external pro-climate stakeholders in the Commission and *high* levels in the Parliament and the *low* levels of involvement in the Council point overall to *medium to high* levels of involvement in the policy process overall.

Recognition of functional interrelations

In the case of the policy process leading to the 2009 RE Directive, there are signs that the functional interrelations between climate policy objectives and RE policy were recognised by policymakers. However, the need for highly ambitious RE policy to achieve *long-term* climate policy objectives was less a part of the discussions.

Climate policy and RE policy interrelate harmoniously, as increasing the share of (most sources of) RE in energy consumption directly results in reductions of GHG emissions from energy, which helps achieve climate policy objectives. Since the agreement in March 2007 in the European Council to move to a 20 per cent target for the share of RE in the EU by 2020, RE policy development was specifically linked to meeting climate policy objectives. In 2007, the European Council stated that it: 'reaffirms the Community's long-term commitment to the EU-wide development of renewable energies beyond 2010' and that it endorses 'a binding target of a 20% share of renewable energies in overall EU energy consumption by 2020' (European Council, 2007: 21). In 2008, the European Council linked this target very clearly to its ambitions to provide leadership in the international climate negotiations, when it stated: 'The EU is committed to maintaining international leadership on climate change and energy ... By delivering on all the targets set by the spring 2007 European Council, the EU will make a major contribution to this objective' (European Council, 2008b: 11). The European Council demonstrates *high* recognition of the interrelations between RE and climate policy, thus outlining the political lines for the various Council formations, at least in the demands for binding policy measures in RE to combat climate change (although it was clear that throughout the policy process, the Council was still the most reticent institutional actor to strengthen policy measures further).

In its proposal, the Commission also clearly outlined its recognition of the functional interrelations between climate policy objectives and RE policy objectives. The RE Directive was proposed by the Commission as part of the climate and energy package that aimed to achieve the so-called 20-20-20 targets by 2020. The opening sentence of the Commission's proposal demonstrates this recognition: 'The Community has long recognised the need to further promote renewable energy given that its exploitation contributes to climate change mitigation through the reduction of greenhouse gas emissions' (European Commission, 2008: 2). The Commission, in its proposals, stuck to the Council's agreed 20 per cent target for RE share in 2020, although recognising the long-term requirements for RE on the road to decarbonisation ought perhaps to have increased the level of ambition. The Commission shows *high* recognition of the functional interrelations between RE policy and climate policy objectives, with RE policy proposals aiming to achieve (short- to medium-term) climate policy objectives.

The Parliament also clearly recognised the functional interrelations between achieving the objectives of climate policy and the objectives of RE policy. Of the three institutions, it was the most ambitious, in that it called for a 25 per cent RE share target for 2020 (European Parliament, 2007). When it could not get this target, the Parliament pushed for strong measures in the RE Directive to ensure the 20 per cent target would be achieved. Although, in the end, it compromised on a number of issues with the Council to ensure agreement was reached, the rapporteur linked the RE Directive throughout the negotiations to the credibility of the EU's leadership on climate change internationally to push for the best results possible (ENDS Europe, 2008k). The Parliament demonstrates *high* recognition of the functional interrelations between RE policy and climate policy objectives throughout the policy process leading to the adoption of the 2009 RE Directive.

In line with Table 2.1, the recognition of the functional interrelations between RE policy and climate policy objectives was *high* in that achieving climate policy objectives is one of the main stated goals of RE policy, and this remained a part of the policy discourse throughout the process. The recognition could not be considered *very high*, however, as policymakers did not seek to be as ambitious on RE policy as long-term climate policy objectives would imply.

In sum, taking the three indicators together, the level of CPI in the policy process can be aggregated as *medium to high*.

CPI in the policy output of the 2009 RE Directive

Figure 3.2 outlines the BAU scenario from 2011 that includes the policy output of the 2009 RE Directive, and compares that to very high levels of CPI on a trajectory from 2000 to 2050. According to this graph, very high levels of CPI in 2020 would see shares of between 37 and 45 per cent of RE in the EU. BAU scenarios from 2006 suggest that the share of RE in the EU was expected to reach about 12 per cent by 2020. Thus, the 2009 RE Directive, with its target to reach a 20 per cent RE share, closes the gap (25 to 33 percentage points) between BAU and very high levels of CPI for 2020 by 8 percentage points. This results in levels of CPI in the policy output of 24 to 32 per cent. According to Table 2.2, this puts the level of CPI in the policy output of the RE Directive at *low* levels.

However, given that the RE Directive was negotiated in 2008, the actual shares of RE in 2005 were known and could form the starting point for a CPI trajectory to 2050 – meaning that the BAU scenario changed between the negotiations on the 2001 RES-E Directive and the 2009 RE Directive, due mainly to deficits in the implementation of previous RE policy. The share of RE in 2005 was close to 9 per cent (EEA, 2008). Starting from 2005 levels, very high levels of CPI in the policy output would reach a share of RE in the EU of between 33 and 40 per cent (see Figure 3.5) in 2020. The 2009 RE Directive can then be considered to close the gap between BAU (12 per cent share in 2020) and very high levels of CPI in 2020 (33 to 40 per cent share), by between 29 and 38 per cent share. This remains at (the higher end of) *low* levels of CPI in the policy output (see Table 2.2).

Finally, the policy output could be measured in a way that tempers the results in the early stages on the path towards decarbonisation. As discussed above on the policy output of the 2001 RES-E Directive, there are some valid arguments to be made about the challenges of developing the RE share in a linear trajectory that makes no distinction between early, costly development, and speedier development in later decades as costs for installing RE fall. Figure 3.4 above suggested a midway point for a very high CPI target of between 17 and 19 per cent RE share in 2010, which corresponds with 2010 levels in the CPI trajectory in Figure 3.5. Measuring from 2005, this implies an increase in the share of RE to 2050 of between 8 and 10 percentage points every five years. This already seems to be a significant increase, considering the history in the EU of achieving an increase of about 1 percentage point every five years since 1990. Nevertheless, it may be realistic to assume that RE shares to 2020 will increase at a slower pace than by 8 to 10 percentage points every five years, due to relatively high upfront costs for

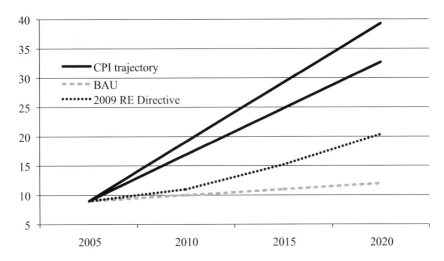

Figure 3.5 CPI in the policy output of the 2009 RE Directive, compared to BAU scenarios and the very high CPI trajectory from 2005 levels

Source: Compiled from (EEA, 2008; European Commission, 2011c), own calculations

expanding RE. An example of reduced early development could rather see between 6 and 7 percentage point increases every 5 years to 2020 – as a direct follow-on from the midway point outlined in Figure 3.4 (with 6 to 7 percentage point increases every five years from 2000 to 2010). Figure 3.6 shows what such a trajectory would look like to 2020 (and beyond to 2050, with steeper RE share development required between 2020 and 2050). Such a scenario suggests that between 9 and 12 percentage point increases in the share of RE would be required every five years from 2020 to 2050.

Following the sort of trajectory outlined in Figure 3.6 implies very high CPI levels of between 27 and 30 per cent RE share by 2020. The output of the RE Directive of a 20 per cent share thus closes the gap between BAU (12 per cent RE share in 2020) and very high levels of CPI by 8 percentage points, with a score of between 44 and 53 per cent. This corresponds to *medium* levels of CPI in the policy output to 2020 for the RE Directive, if we assume that steeper increases of RE in the overall share of EU energy can be more easily achieved after 2020. Such a trajectory does, however, leave a substantial part of the work for the later decades leading to 2050, while climate scientists and scholars argue in favour of policies that ensure that GHG emissions peak as soon as possible (IPCC, 2007, 2013).

Overall, the level of CPI in the policy output of the 2009 RE Directive can be considered *low to medium*, but closer to *low* if it is assumed that linear or early action is required to achieve climate policy objectives.

88 Renewable energy

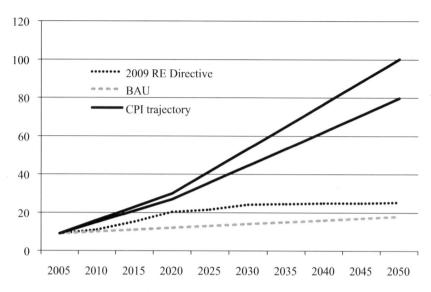

Figure 3.6 CPI trajectory for the RE share to 2050, with lower expectations to 2020
Source: Compiled from (EEA, 2008; European Commission, 2011c), own calculations

Summary and outlook to 2030 and beyond

In this chapter, I examined the level of CPI in the policy output and process of the 2001 RES-E and 2009 RE Directives. The 2001 RES-E Directive aimed to increase the share of RE in EU energy consumption to 12 per cent by 2010 (corresponding to an increase in the share of RE in electricity generation to about 22 per cent by 2010). The targets agreed under the 2001 RES-E Directive were indicative targets, and reports indicate that they were not met (EEA, 2008; European Commission, 2006a, 2011c). The 2001 Directive was proposed with the main motivation of achieving part of the EU's commitment to reduce its GHG emissions under the Kyoto Protocol. The 2009 RE Directive was proposed in 2008 as part of the climate and energy package. It provided the legislative framework for the EU to meet the target of increasing the share of RE in EU energy consumption to 20 per cent by 2020, as endorsed by the European Council in March 2007 (European Council, 2007). The 2009 Directive imposed binding targets on member states to achieve the overall 20 per cent target.

Both the 2001 and 2009 Directives were proposed with part of their aims being the achievement of climate policy objectives. In the case of the 2001 RES-E Directive, this was framed in terms of contributing to achieve the EU's emission reduction commitments under the Kyoto Protocol – to reduce GHG emissions by 8 per cent between 2008 and 2012 compared to 1990 levels. This can be considered a short-term climate policy objective. The longer-term climate policy objectives (to meet the 2050 goals or to limit global temperature increase to 2° Celsius) did not seem to play a major role in policy discussions leading to the adoption of the RES-E Directive. For the 2009 RE Directive, the EU aimed to demonstrate through unilateral action that it was a credible leader on climate change. It agreed the 2009 RE Directive in the first reading, motivated to influence international climate negotiations in 2009. The EU RE policy therefore has been closely linked to policy objectives on climate change between 2000 and 2010. At first glance, then, RE policy can be considered a case of well-integrated climate and energy policy.

As the analysis in this chapter revealed, however, the levels of CPI in EU RE policy are far from ideal in the policy output, when measured against benchmarks that aim to achieve the goals of long-term climate policy. In both the 2001 and 2009 Directives, CPI in the policy output was, at best, *low to medium*. There was limited improvement over the course of the decade, and the poor implementation of the 2001 RES-E Directive meant later policy developments were catching-up on potential improvements that did not materialise. The level of CPI in the policy process was *low to medium* in the 2001 Directive, but this level had increased to *medium to high* during the negotiations on the 2009 Directive, with more guarantees through procedures for the involvement of pro-climate stakeholders in the process.

The next stage of EU RE policy development is the 2030 climate and energy framework, under discussion at the time of writing (2014–2015). In October 2014, the European Council accepted a target to increase RE share in the EU28 to at least 27 per cent by 2030, but without agreeing on the need for mandatory member state targets to achieve the overall objective (European Council, 2014: 5). At the time of writing, the Commission has yet to come forward with policy proposals for implementing the agreed 2030 objective, but it is clear that an increase of 7 percentage points in 10 years (between 2020 and 2030) is a lower increase than one could expect from a long-term trajectory to 2050 (see Figure 3.6, for example). As the timeframe to 2050 shortens, the EU has still not succeeded in advancing policy proposals that integrate the long-term climate policy objective to reduce GHG emissions by 80–95 per cent by 2050 (Dupont and Oberthür, 2015).

Note

1   The 2011 BAU scenario here represents an average of the five BAU scenarios presented by the Commission in the 2011 impact assessment to the energy roadmap. These scenarios vary according to price of imported energy and GDP. The lowest expected share of RE in 2050 is 23 per cent – expected in a BAU scenario with low energy import prices. The highest expected share of RE in 2050 is 27 per cent – from the BAU scenario with high-energy import price assumptions (European Commission, 2011b).

References

Boasson, E. L. and Wettestad, J. (2013). *EU Climate Policy: Industry, Policy Innovation and External Environment*. Farnham: Ashgate.
Council of the European Union. (1986). Resolution Concerning New Community Energy Policy Objectives for 1995 and Convergence of the Policies of the Member States. *15/16.IX.86*.
Council of the European Union. (1996). 1939th Environment Council Meeting. *8518/96 (Presse 188)*.
Council of the European Union. (2001). Common Position Adopted by the Council on 23 March 2001 with a View to the Adoption of Directive of the European Parliament and of the Council on the Promotion of Electricity Produced from Renewable Energy Sources in the Internal Electricity Market. *5583/1/01*.
Council of the European Union. (2008a). 2875th Council Meeting: Transport, Telecommunications and Energy. *10310/08 (Presse 162)*.
Council of the European Union. (2008b). 2895th Council Meeting: Transport, Telecommunications and Energy. *13649/08 (Presse 276)*.
Council of the European Union. (2008c). 2898th Council Meeting: Environment. *13857/08 (Presse 282)*.
Council of the European Union. (2008d). 2913rd Council Meeting: Transport, Telecommunications and Energy. *16920/08 (Presse 362)*.
Council of the European Union. (2008e). Preparation of the Council (Environment) Meeting on 5 June 2008 and of the TTE Council (Energy) Meeting on 6 June 2008. *9648/08*.
DG Climate Action. (2014). *Support to the Review of Directive 2009/31/EC on the Geological Storage of Carbon Dioxide (CCS Directive)*. Luxembourg: Publications Office of the European Union.
Dupont, C. and Oberthür, S. (2012). Insufficient Climate Policy Integration in EU Energy Policy: the Importance of the Long-Term Perspective. *Journal of Contemporary European Research*, 8(2), 228–247.
Dupont, C. and Oberthür, S. (2015). The European Union. In E. Lövbrand and K. Bäckstrand (eds), *Research Handbook on Climate Governance* (Forthcoming). Cheltenham: Edward Elgar.
ECF. (2010). *Roadmap 2050. A Practical Guide to a Prosperous, Low Carbon Europe*. Brussels: European Climate Foundation.
EEA. (2008). *Energy and Environment Report 2008*. Copenhagen: European Environment Agency.
EEA. (2011). *Renewable Energy 2000 to 2010 – from Toddler to Teen*. Copenhagen: European Environment Agency.
EEA. (2014). *Trends and Projections in Europe 2014. Tracking Progress Towards Europe's Climate and Energy Targets for 2020*. Copenhagen: European Environment Agency.
ENDS Europe. (1999a). EU Commission Attacked Over Renewables Law, *22 October 1999*.
ENDS Europe. (1999b). Plan for EU Renewable Energy Law Shelved, *9 February 1999*.
ENDS Europe. (1999c). Strong EU Renewable Energy Law Demanded, *28 April 1999*.
ENDS Europe. (2000a). EU Renewables Law Gets Ministerial Approval, *7 December 2000*.
ENDS Europe. (2000b). France Aims for Quick Deal on EU Renewables, *31 May 2000*.
ENDS Europe. (2000c). MEPs Demand Binding Renewables Targets, *17 November 2000*.
ENDS Europe. (2000d). Slow Going for EU Renewables Directive, *8 November 2000*.

ENDS Europe. (2001a). EU Renewable Energy Law all but Finalised, *20 June 2001*.
ENDS Europe. (2001b). MEP Stands by Legal Renewable Energy Targets, *6 March 2001*.
ENDS Europe. (2008a). Council Set to Reject EU-Wide Renewables Trading, *27 June 2008*.
ENDS Europe. (2008b). Crisis Leaves the Climate Glass Half Empty, *19 December 2008*.
ENDS Europe. (2008c). EU Renewable Energy Law Deal Confirmed, *9 December 2008*.
ENDS Europe. (2008d). EU States Reach Accord on Renewables Plan, *29 October 2008*.
ENDS Europe. (2008e). European Leaders Pass Climate "Credibility Test," *12 December 2008*.
ENDS Europe. (2008f). MEP to Propose Big Changes to Renewable Plans, *13 May 2008*.
ENDS Europe. (2008g). MEPs Agree to Compromise on EU Renewables Vote, *10 September 2008*.
ENDS Europe. (2008h). MEPs Back Beefed-Up Renewable Energy Law, *11 September 2008*.
ENDS Europe. (2008i). New EU Renewable Energy Law "Finalised", *2 December 2008*.
ENDS Europe. (2008j). Rapporteur Kicks off EU Renewables Law Debate, *31 January 2008*.
ENDS Europe. (2008k). EU Renewable Energy Law Deal Confirmed. *09 December 2008*.
EREC. (2010). *RE-thinking 2050: a 100% Renewable Energy Vision for the European Union.* Brussels: European Renewable Energy Council.
EREC and Greenpeace. (2010). *Energy [R]evolution: Towards a Fully Renewable Energy Supply in the EU 27.* Brussels: Greenpeace International and European Renewable Energy Council.
European Commission. (1992). Specific Actions for Greater Penetration for Renewable Energy Sources: ALTENER. COM(92) 180.
European Commission. (1996). Energy for the Future: Renewable Sources of Energy. Green Paper for a Community Strategy. COM(1996) 576.
European Commission. (1997). Energy for the Future: Renewable Sources of Energy. White Paper for a Community Strategy and Action Plan. COM(1997) 599.
European Commission. (2000a). Amended Proposal for a Directive of the European Parliament and of the Council on the Promotion of Electricity from Renewable Energy Sources in the Internal Electricity Market. COM(2000) 884.
European Commission. (2000b). Proposal for a Directive of the European Parliament and of the Council on the Promotion of Electricity from Renewable Energy Sources in the Internal Electricity Market. COM(2000) 279.
European Commission. (2002). Towards a Reinforced Culture of Consultation and Dialogue – General Principles and Minimum Standards for Consultation of Interested Parties by the Commission. COM(2002) 704.
European Commission. (2006a). Green Paper Follow-Up Action: Report on Progress in Renewable Energy. COM(2006) 849.
European Commission. (2006b). Green Paper: a European Strategy for Sustainable, Competitive and Secure Energy. COM(2006) 105.
European Commission. (2006c). Summary Report on the Analysis of the Debate on the Green Paper "a European Strategy for Sustainable, Competitive and Secure Energy". SEC(2006) 1500.
European Commission. (2007a). Commission Staff Working Document Accompanying Document to the Renewable Energy Roadmap. Renewable Energies in the 21st Century: Building a More Sustainable Future. Impact Assessment. SEC(2006) 1719.

European Commission. (2007b). Communication from the Commission: Renewable Energy Roadmap. Renewable Energies in the 21st Century: Building a More Sustainable Future. COM(2006) 848.
European Commission. (2007c). Energy for a Changing World – an Energy Policy for Europe. COM(2007) 1.
European Commission. (2007d). Limiting Global Climate Change to 2 Degrees Celsius. The Way Ahead for 2020 and Beyond. COM(2007) 2.
European Commission. (2008). Proposal for a Directive of the European Parliament and of the Council on the Promotion of the Use of Energy from Renewable Sources. COM(2008) 19.
European Commission. (2011a). Communication from the Commission: Energy Roadmap 2050. COM(2011) 885/2.
European Commission. (2011b). Communication from the Commission. A Roadmap for Moving to a Competitive Low Carbon Economy in 2050. COM(2011) 112.
European Commission. (2011c). Impact Assessment Accompanying the Document: Energy Roadmap 2050. SEC(2011) 1565 Part Two.
European Council. (2006). *Presidency Conclusions*, March 2006. Brussels: Council of the European Union.
European Council. (2007). *Presidency Conclusions*, March 2007. Brussels: Council of the European Union.
European Council. (2008a). *Presidency Conclusions*, December 2008. Brussels: Council of the European Union.
European Council. (2008b). *Presidency Conclusions*, March 2008. Brussels: Council of the European Union.
European Council. (2009). *Presidency Conclusions*, October 2009. Brussels: Council of the European Union.
European Council. (2014). *Conclusions*, Document EUCO 169/14, October 2014. Brussels: European Council.
European Parliament. (2000a). European Parliament First Reading on the Proposal for a European Parliament and Council Directive on the Promotion of Electricity from Renewable Energy Sources in the Internal Electricity Market. T5-0514/2000.
European Parliament. (2000b). Report on the Proposal for a European Parliament and Council Directive on the Promotion of Electricity from Renewable Sources in the Internal Electricity Market. A5-0320/2000.
European Parliament. (2001). Recommendation for Second Reading on the Council Common Position for Adopting a European Parliament and Council Directive on the Promotion of Electricity Produced from Renewable Energy Sources in the Internal Electricity Market. A5-0227/2001.
European Parliament. (2006). European Parliament Resolution on a European Strategy for Sustainable, Competitive and Secure Energy. 2006/2113 (INI).
European Parliament. (2007). European Parliament Resolution on Climate Change. P6_TA(2007)0038.
Eurostat. (2015). Eurostat Online Database. Available at: ec.europa.eu/eurostat/data/database.
Hauser, H. (2011). European Union Lobbying Post Lisbon: an Economic Analysis. *Berkeley Journal of International Law*, 29(2), 680–709.
Haverland, M. and Romeijn, M. (2007). Do Member States Make European Policies Work? Analysing the EU Transposition Deficit. *Public Administration*, 85(3), 757–778.

Heaps, C., Erickson, P., Kartha, S. and Kemp-Benedict, E. (2009). *Europe's Share of the Climate Challenge: Domestic Actions and International Obligations to Protect the Planet.* Stockholm: Stockholm Environment Institute.

Hildingsson, R., Stripple, J. and Jordan, A. (2010). Renewable Energies: a Continuing Balancing Act? In A. Jordan, D. Huitema, H. van Asselt, T. Rayner and F. Berkhout (eds), *Climate Change Policy in the European Union: Confronting the Dilemmas of Mitigation and Adaptation?* (pp. 103–124). Cambridge: Cambridge University Press.

Howes, T. (2010). The EU's New Renewable Energy Directive (2009/28/EC). In S. Oberthür and M. Pallemaerts (eds), *The New Climate Policies of the European Union: Internal Legislation and Climate Diplomacy* (pp. 117–150). Brussels: VUB Press.

IPCC. (2007). *Climate Change 2007. Fourth Assessment Report: Synthesis Report.* Geneva: Intergovernmental Panel on Climate Change.

IPCC. (2013). Summary for Policymakers. In T. F. Stoker, D. Qin, G.-K. Plattner, M. Tignor, S. K. Allen, J. Boschung, ... P. M. Midgley (eds), *Climate Change 2013: the Physical Science Basis. Contribution of Working Group I to the Fifth Assessment Report of the Intergovernmental Panel on Climate Change.* Cambridge: Cambridge University Press.

Jansen, J. C. and Uyterlinde, M. A. (2004). A Fragmented Market on the Way to Harmonisation? EU Policy-Making on Renewable Energy Promotion. *Energy for Sustainable Development, 8*(1), 93–107.

Lampinen, R. and Uusikylä, P. (1998). Implementation Deficit – Why Member States Do Not Comply with EU Directives. *Scandinavian Political Studies, 21*(3), 231–251.

Lauber, V. (2001). The Different Concepts of Promoting RES-Electricity and Their Political Careers. In F. Biermann, R. Brohm and K. Dingwerth (eds), *2001 Berlin Conference on the Human Dimensions of Global Environmental Change, "Global Environmetal Change and the Nation State"* (pp. 296–304). Berlin: Potsdam Institute for Climate Impact Research (PIK).

Lenschow, A. (2002). Greening the European Union: an Introduction. In A. Lenschow (ed.), *Environmental Policy Integration: Greening Sectoral Policies in Europe* (pp. 3–21). London: Earthscan.

Moomaw, W., Yamba, F., Kamimoto, M., Maurice, L., Nyboer, J., Urama, K. and Weir, T. (2011). Introduction. In O. Edenhofer, R. Pichs-Madruga, Y. Sokona, K. Seyboth, P. Matschoss, S. Kadner, ... C. von Stechow (eds), *IPCC Special Report on Renewable Energy Sources and Climate Change Mitigation.* Cambridge and New York: Cambridge University Press.

Nilsson, M. (2011). EU Renewable Electricity Policy: Mixed Emotions Towards Harmonization. In V. L. Birchfield and J. S. Duffield (eds), *Towards a Common European Union Energy Policy: Problems, Progress, and Prospects* (pp. 113–130). New York: Palgrave Macmillan.

Nilsson, M., Nilsson, L. J. and Ericsson, K. (2009). The Rise and Fall of GO Trading in European Renewable Energy Policy: the Role of Advocacy and Policy Framing. *Energy Policy, 37*(2009), 4454–4462.

Oberthür, S. and Pallemaerts, M. (2010a). The EU's Internal and External Climate Policies: an Historical Overview. In S. Oberthür and M. Pallemaerts (eds), *The New Climate Policies of the European Union: Internal Legislation and Climate Diplomacy* (pp. 27–63). Brussels: VUB Press.

Oberthür, S. and Pallemaerts, M. (eds) (2010b). *The New Climate Policies of the European Union: Internal Legislation and Climate Diplomacy.* Brussels: VUB Press.

Odenberger, M. and Johnsson, F. (2010). Pathways for the European Electricity Supply System to 2050: the Role of CCS to Meet Stringent CO_2 Reduction Targets. *International Journal of Greenhouse Gas Control, 4*(2), 327–340.

PricewaterhouseCoopers. (2010). *100% Renewable Electricity. A Roadmap to 2050 for Europe and North Africa*. UK: PricewaterhouseCoopers.

Reichardt, K., Pfluger, B., Schleich, J. and Marth, H. (2012). With or Without CCS? Decarbonising the EU Power Sector. *Responses Policy Update*, 3(July 2012).

Rowlands, I. H. (2005). The European Directive on Renewable Electricity: Conflicts and Compromises. *Energy Policy, 33*, 965–974.

WWF. (2011). *The Energy Report: 100% Renewable Energy by 2050*. Gland, Switzerland: World Wide Fund for Nature.

4 EU policy on the energy performance of buildings

In this chapter, I discuss the second case study – EU policy on the energy performance of buildings (EPB). Here, I turn attention to policies to manage and reduce energy consumption. With buildings representing the largest consumer of energy in the EU (followed by transport and industry; European Commission, 2013: 6), any policy to reduce energy consumption must deal with the buildings sector. I analyse the level of CPI in the policy processes and outputs of legislative measures to improve the energy performance of buildings, with a particular focus on the EU's 2002 Energy Performance of Buildings Directive (EPBD, Directive 2002/91/EC) and its 2010 recast (Directive 2010/31/EU). I set the discussion on these two Directives within the broader context of EU energy efficiency policy development over time and possible future policy developments.

Energy efficiency in the EU

Improving energy efficiency can be considered a win-win solution for a wide range of issues, including the goals of EU energy policy more broadly: security of energy supply, competitiveness and sustainable use of energy. Defined as output divided by total input, or as actions that are designed to save fuels (EEA, 2013), energy efficiency involves reducing the amount of energy consumed. This implies reducing energy consumption of products, buildings, transport, power generation and heating. While several actions can be taken at the demand side (for example, consumers choose the most efficient products and appliances, reduce their home heating temperature and electricity consumption), other actions are stimulated at the supply side (e.g. efficiency measures in power generation). Energy efficiency policy, thus, is not usually an end in itself, but rather a means to achieve other overarching policy objectives (Boasson and Dupont, 2015).

Energy efficiency is often measured in terms of the 'energy intensity' of an economy. The less energy intensive an economy (in other words, the less energy consumed for a fixed cost), the more efficient the economy is. As Figure 4.1 shows, the energy intensity of the EU28 has declined over time, indicating energy efficiency improvements.

Levels of energy consumption, however, have rather fluctuated over time, and not followed a consistent downward trend. As Figure 4.2 shows, the biggest drop

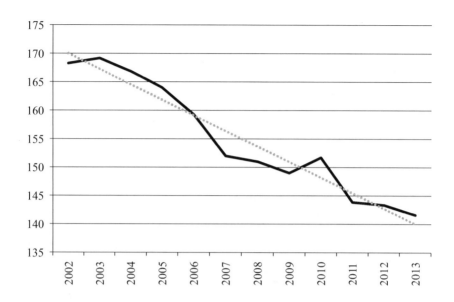

Figure 4.1 Development of energy intensity of the economy of the EU28 between 2002 and 2013, measured in kilograms of oil equivalent per €1000, with (dotted) trend line

Source: Compiled from (Eurostat, 2015)

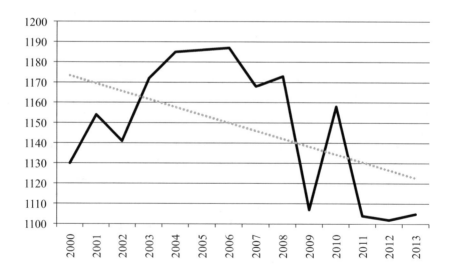

Figure 4.2 Final energy consumption in the EU28 from 2000 to 2013, measured in mega tonnes of oil equivalent (Mtoe), with (dotted) trend line

Source: Compiled from (Eurostat, 2015)

in energy consumption in the EU in the 2000s occurred between 2008 and 2009, with the onset of the economic crisis, but energy consumption increased again between 2009 and 2010. Although energy efficiency has improved overall in the EU, these improvements in efficiency have sometimes been compensated by increases in energy consumption generally.

These figures demonstrate one of the major difficulties with energy efficiency measures, known as the rebound effect. The rebound effect 'refers to behavioural changes or other systemic responses that can partly or fully offset the beneficial environmental effects of … new technologies' (EEA, 2012: 22). As energy efficiency measures lead to reduction in energy bills, consumers use the saved money to buy even more or bigger energy-consuming appliances. The rebound effect may have economy-wide equilibrium effects: as efficiency improves, consumption increases, thus resulting in no net benefits for the economy. However, other factors also play a role, such as increases in income and increases in population. For policymakers, promoting behavioural change and taking account of this 'rebound effect' represent challenges when designing measures to improve energy efficiency. Nevertheless, there seems to be a particular tipping point in energy efficiency improvements beyond which the rebound effect is less problematic. As more and more products, appliances, buildings, and so on, have high levels of energy efficiency, the less important the rebound effect becomes.

The EU has put in place an array of policy measures aimed at improving energy efficiency. Table 4.1 lists some of the main legislative measures adopted. As far back as 16 September 1986, the Council set the objective of improving energy efficiency by 20 per cent by 1995 compared to 1985 levels (Council of the European Union, 1986). At this time, it was clear that environmental or climate objectives did not play any role in motivating energy efficiency policy. The Council's 1986 resolution highlighted energy security (especially in the context of rising EU dependence on oil imports) as the main motivation for improving energy efficiency.

The Council's position in support of energy efficiency for energy security reasons is a direct consequence and follow-on from the oil crises of the 1970s. Improving energy efficiency and thus reducing demand for oil was seen as an effective response to concerns about growing dependence on foreign energy imports, and thus an effective policy measure for promoting energy security in the EU (Boasson and Dupont, 2015; Boasson and Wettestad, 2013; European Commission, 1979a, 1984). Several member states responded to the oil crises with policy measures aimed at improving energy efficiency in their building sectors. National level energy performance standards for new buildings were put in place in the Netherlands, Denmark, Germany and France (European Commission, 1979b: 2). The 1979 report from the Commission on the EU's early programme for energy saving highlighted the future strategy for improving energy efficiency that later policy would follow, stating that 'future savings will increasingly require investment in new equipment or buildings, or retrofitting the old, and more energy conscious behaviour from both investors and consumers' (European Commission, 1979b: 1).

The Commission continued to suggest energy efficiency measures in the 1980s. Its 1984 proposal on 'a European policy for the rational use of energy in the building sector' highlighted the high level of energy consumption of buildings, and described the potential for the buildings sector to save approximately 50 per cent energy consumption of the total saving potential of all sectors (European Commission, 1984: 1). The strategy for improving the energy performance of buildings included: the promotion of thermal auditing of buildings; technical improvements and regulations; use of funds for supporting efficiency measures; consumer/energy-user information. Much of the future regulation on energy efficiency, whether for buildings or appliances, incorporated these same themes. It was not until the late 1980s, however, that concrete policy measures at the EU-level were introduced (rather than simply discussing potential targets for member states). The 1989 Construction Products Directive was a first move at the EU level into building sector legislation (see Table 4.1).

As the energy efficiency issue moved down the political agenda, with the memory of the 1970s oil crises fading, the Commission nevertheless continued to propose (indicative) targets and policies in energy efficiency to enhance energy security, while pointing out insufficient progress on energy efficiency. In 1991, the Commission noted that the objective to improve energy efficiency by 20 per cent by 1995 compared to 1985 'was unlikely to be attained unless stronger policy measures were taken' (European Commission, 1991: 3). Much of the EU's energy efficiency policy in the 1990s was still considerably watered down from the Commission's original proposal by member states during the policy negotiations. Member states were unconvinced about the importance of EU energy efficiency policy measures, and discussions around subsidiarity surfaced repeatedly. Later policy failures were attributed to inadequate implementation by member states (Boasson and Dupont, 2015; Boasson and Wettestad, 2013; Henningsen, 2011).

Table 4.1 Some of the main pieces of EU legislation promoting energy efficiency

Year	Legislation	Code
1989	Construction products Directive	89/106/EEC
1992	Energy labelling Directive	92/75/EEC
1992	Efficiency for hot-water boilers Directive	92/42/EEC
1993	SAVE Directive	93/76/EEC
2002	Energy performance of buildings Directive	2002/91/EC
2004	Promotion of cogeneration Directive	2004/8/EC
2005	Ecodesign Directive	2005/32/EEC
2006	Energy services Directive	2006/32/EC
2009	Ecodesign Directive – recast	2009/125/EC
2010	Energy labelling Directive – recast	2010/30/EU
2010	Energy performance of buildings Directive – recast	2010/31/EU
2012	Energy efficiency Directive	2012/27/EU

The Commission began linking energy efficiency policy measures to the need to combat climate change in the 1990s. Boasson and Wettestad describe how the international climate negotiations provided an opportunity for the Commission to push for ambitious energy efficiency measures internally in the EU (2013: 152). Combating climate change became an added reason to pursue energy efficiency policy (European Commission, 1992, 1998). Since then, energy efficiency policy measures have come to be considered as *climate policies* as well as *energy policies*, and, at first glance at least, could be considered as excellent examples of climate policy integration into energy policy. It is certainly difficult to conceive of efficiency measures that could be *bad* for combating climate change. With the energy sector responsible for 80 per cent of GHG emissions in the EU (CANEurope, 2011: 14; EEA, 2010: 31), any reduction in the consumption of energy is beneficial for achieving a reduction of GHG emissions.

Energy performance of buildings in the EU

In 2010, the Buildings Performance Institute Europe (BPIE) carried out the first comprehensive study of the state of the then EU27's building stock (BPIE, 2011). The study showed the heterogeneity of the EU's buildings: different member states have buildings of different types, ages and energy performance. Some of the overarching statistics for the EU are nonetheless telling of the enormous challenge for EU policy to improve the energy performance of buildings.

While 25 per cent of the total of the EU's building stock is non-residential (offices, retail, wholesale, public buildings, etc.), the remaining 75 per cent residential buildings are divided between owner-occupied and rented residences, with great variations among member states. In 2009, 42 per cent of the EU27 population lived in apartments, 34 per cent in detached houses and 23 per cent in semi-detached or terraced houses. Additionally, the energy consumption of the building sector has increased steadily since 1990 (by about 1 per cent per year; European Commission, 2013: 6), although the increase in energy consumption was greatest in non-residential buildings. In general, trends in energy consumption in buildings are marked by a decrease in the consumption of solid fuels and oil and an increase in the use of natural gas and electricity and in overall energy consumption (BPIE, 2011; European Commission, 2013).

Reducing the energy consumption of buildings involves targeting the main sources of energy consumption, namely space heating (and cooling), water heating and the use of electric and electronic appliances (EEA, 2012: 32). Therefore, energy performance standards for buildings include measures for making buildings more secure against outside cold or hot air, through insulation in the building envelope (walls, roof, windows), combined with the latest airing technologies. As energy performance standards gradually became more ambitious, the general efficiency of new buildings has increased. One major challenge, however, remains the improvement of the energy performance of the existing building stock.

According to BPIE's report, about 40 per cent of the EU27's building stock was built before 1960, when energy regulations were very limited. Less than 20 per cent of EU buildings were built between 1990 and 2010 (BPIE, 2011: 9–10). This is crucial when considering the potential for advancing improvements in the energy performance of buildings in the EU – policies need to tackle measures in the existing building stock to impact the energy consumption of buildings. Thus, it is important also to take note of the rates of building renovations. As of 2013, the renovation rates for buildings in the EU were about 1 to 2 per cent per year (European Commission, 2013). The type of renovations being undertaken is also crucial for effective reduction in energy consumption in buildings. BPIE proposes that renovations leading to a 'nearly-zero energy building' hold the most promise for reducing energy consumption of buildings by 2050. This type of renovation includes:

> the wholesale replacement or upgrade of all elements which have a bearing on energy use, as well as the installation of renewable energy technologies in order to reduce energy consumption and carbon emission levels to close to zero, or, in the case of an 'energy positive' building, to less than zero (i.e. a building that produces more energy from renewable sources than it consumes over an annual cycle). The reduction of the energy needs towards very low energy levels (i.e. passive house standards, below 15kWh/m^2 [kilo Watt hours per metre squared] per year) will lead to the avoidance of a traditional heating system ...
>
> (BPIE, 2011: 103)

With such a type of renovation, BPIE argues that to have any chance of meeting 2050 targets for buildings, the entire building stock (100 per cent) of the EU will need to be renovated between 2010 and 2050 (a renovation rate of 2.5 per cent per year, with renovations meeting the nearly-zero energy building standard) (2011: 109). These are important figures to note for measuring the level of CPI in the policy process and output of the EU's EPB policy.

The less (fossil fuel-generated) energy consumed in buildings, the less GHG emissions from energy production and use. Analyses of how the EU will achieve its long-term climate policy objective of reducing GHG emissions in the EU by between 80 and 95 per cent by 2050 all point to the substantial role to be played by energy efficiency measures (see, for example, CANEurope, 2011; ECF, 2010; European Commission, 2011a; Heaps, Erickson, Kartha and Kemp-Benedict, 2009; IEA, 2011). Some estimate that energy efficiency measures can achieve about half of the 80–95 per cent emissions reduction goal by 2050 (Wesselink, Harmsen and Eichhammer, 2010). With buildings accounting for approximately 40 per cent of the EU's total energy consumption, measures to improve the efficiency of the economy must include energy performance measures for the buildings sector.

The complexity of improving the energy performance of buildings, however – due to the climatic variances across the EU, the ensuing differences in building standards, and the fact that the building sector is largely dominated by small businesses operating at a local level – represents a major challenge to agreement on and implementation of energy performance measures in buildings (Boasson and Dupont, 2015). These barriers are linked to what is considered a 'market failure' in terms of the promotion of energy efficiency (Dawson and Spannagle, 2009; Richter, 2010; Roosa, 2010). Lack of awareness about the benefits of investing in energy performance improvements, split incentives among constructors, owners and tenants, and insufficient professional training all hamper the uptake of energy efficiency improvements for buildings generally (Boasson and Dupont, 2015; European Commission, 2011b). Additionally, there are certain barriers to introducing EU-level policy on the energy performance of buildings, such as subsidiarity rules and the disparate construction industry that is generally made up of small to medium businesses operating in a locally-defined geographic area (with limited organised representation at the EU level; Boasson and Dupont, 2015; Boasson and Wettestad, 2013).

CPI into EU energy performance of buildings policy

Considering the challenges discussed above about how to make effective policy to improve the energy performance of buildings at EU level, integrating long-term climate policy objectives may provide some direction and motivation for policy development. As with the EU's RE policy, we can expect very high levels of CPI in the policy process if pro-climate stakeholders internally in the EU institutions and externally to the decision-making process have a strong voice in the policy discussions. In addition, if policymakers recognise the functional interrelations between policy to improve the energy performance of buildings and long-term climate policy objectives there are higher opportunities for integrating climate policy into buildings policy.

Whether the level of CPI throughout the process is then evident in the policy output depends on measuring the policy output against a benchmark policy output that would be in line with 2050 climate policy objectives. In the buildings sector, a benchmark level of improvements that is in line with climate policy objectives would imply significant improvements in the energy performance of the EU's entire building stock. Very high CPI would then lead to decreases in the energy consumption of the entire EU building stock by 2050 in line with targets for decarbonisation. Importantly, such a benchmark implies that policy outputs must deal simultaneously with putting in place energy performance standards for new buildings and for existing buildings through renovations.

While scenarios towards decarbonisation by 2050 show the great role to be played by improved energy efficiency measures, few scenarios outline studies on the contribution of the buildings sector in particular. Achieving the energy efficiency improvements generally are seen to require retrofitting the current building stock at a faster pace and to high-energy performance standards, and of ensuring new builds quickly conform to high-energy performance building standards, such as the passive building, zero-energy or energy-plus buildings (BPIE, 2011; EREC, 2010; Heaps et al., 2009). In a system with 100 per cent of the EU's building stock reaching very high standards of energy performance, it may be possible to conceive of reducing overall energy consumption in buildings to zero. However, in no scenario to 2050 is this considered a feasible option. This is for two main reasons. First, the rate of retrofitting the existing building stock in the EU will have to increase dramatically from the 1 to 2 per cent rate recorded in 2013 (BPIE, 2011; European Commission, 2013: 7). The type of renovations will also have to ensure a very high standard of energy performance in buildings by 2050 to reduce energy consumption in buildings to as close to zero as possible. Second, there are several historical buildings and monuments that simply cannot be retrofitted to meet high-energy performance standards of buildings.

Therefore, reasonable estimates show that the buildings sector can reduce its energy consumption to a maximum level of 200 Mtoe in the EU in 2050 (compared to about 460 Mtoe in 2010; European Commission, 2011c) with a retrofitting rate of between 2.5 and 5 per cent, and achieving energy efficiency standards (e.g. a mix of 'nearly-zero energy buildings' and 'energy plus' buildings) in about 90 per cent of the EU's building stock (BPIE, 2011; Heaps et al., 2009). With a maximum 200 Mtoe of energy consumption in the buildings sector, a reasonable very high level of CPI could include energy consumption in buildings of between 100 (or less) and 200 Mtoe. Such a range allows for the continued higher levels of consumption in historical buildings, but assumes an aggressive push to retrofit the existing building stock. Figure 4.3 outlines what such a

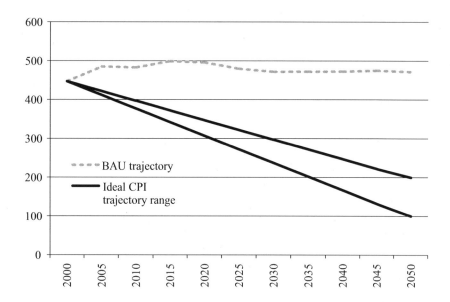

Figure 4.3 Very high CPI trajectory for energy consumption in buildings in EU28, from 2000 to 2050, compared to the BAU trajectory (that includes actual levels of consumption from 2000 to 2010; measured in Mtoe)

Source: Compiled from (European Commission, 2011; Heaps, Erickson, Kartha and Kemp-Benedict, 2009), own calculations

decrease in energy consumption would look like from 2000, and compares this very high level of CPI with the European Commission's (2011) BAU scenario to 2050, which shows energy consumption of buildings at 472 Mtoe in 2050. Very high levels of CPI are dramatically lower than BAU scenarios, so policy measures need to close a significant gap.

Many of the early pieces of energy efficiency legislation (see Table 4.1) were relevant for improving the energy performance of buildings. The 1989 construction products Directive highlighted that construction materials and products should promote energy efficiency. The 1992 labelling Directive focused specifically on the provision of energy consumption information on household appliances (Directive 92/75/EEC). The 1992 hot-water boilers Directive (92/42/EEC) determined the essential efficiency requirements for hot-water boilers, which then receive the 'CE' marker indicating their conformity with (all applicable) legislation. The SAVE (Specific Actions for Vigorous Energy Efficiency) programme of the early 1990s provided financial support for actions to improve energy efficiency (see Council Decision 91/565/EEC). The 1993 SAVE Directive aimed at limiting carbon dioxide emissions by improving energy efficiency, notably through programmes in the fields of energy certification of buildings; thermal insulation of new buildings; regular inspection of boilers; among others (Art. 1). But it did not include specific commitments for member states. With hindsight, the success of the SAVE programme (and many other early energy efficiency measures) was hampered by the lack of member state ambition and the limited financing available to implement its objectives (Boasson and Dupont, 2015; Boasson and Wettestad, 2013; Oberthür and Roche Kelly, 2008).

Against the background of failure to achieve energy efficiency targets and weak policy measures, it became clear that further legislation would be required to spur improvements in energy efficiency (Henningsen, 2011). The 2002 EPBD and its 2010 recast represent the main pieces of legislation on buildings under examination in this chapter. These pieces of legislation require member states to adopt technological measures to improve the energy quality of buildings. The main elements of the Directives include: a method for calculating the energy performance of buildings; requirements for national building codes; requirements on the control of heating and cooling systems in buildings; recommendations for financial support measures; and outlines for national energy certification of buildings (Boasson and Dupont, 2015; Boasson and Wettestad, 2013: 130; Directives 2002/91/EC and 2010/31/EU).

Energy Performance of Buildings Directive 2002/91/EC

The 2001 legislative proposal for a Directive on the energy performance of buildings was drafted by DG Energy under the leadership of Commissioner Loyola de Palacio (European Commission, 2001). This proposal was made under Article 175 of the Treaty establishing the European Community (today's Article 192 of the Treaty on the Functioning of the EU, TFEU), placing the proposal under the legal competence of the environmental chapter of the Treaty. Climate change and the EU's commitments to reduce its GHG emissions are cited in the proposal as major rationales for taking action on improving the energy performance of buildings (European Commission, 2001: 15, 19). The proposal set out minimum standards for energy performance and included the following suggested requirements:

a) A general framework for a common methodology to calculate the energy performance of buildings;
b) The application of minimum energy performance standards in new buildings;
c) The application of minimum energy performance standards in large existing buildings that are to undergo renovation;
d) The energy certification of buildings, and the display of this energy information in public buildings, and;
e) The regular inspection of boilers, air-conditioning and heating systems in existing buildings.

(pp. 21–22)

The proposal moved into first reading in the Parliament on 17 May 2001. The ITRE committee, with Spanish MEP Aledo Vidal-Quadras of the European People's Party (EPP) group as rapporteur, was responsible for drafting the first reading opinion. Cristina García-Orcoyen Tormo, also a member of the EPP group, acted as rapporteur for providing the opinion of the ENVI Committee. The first reading committee report was brought to the Parliament's plenary on 18 December 2001, with debate in the plenary taking place on 4 February the following year, and the adoption of the opinion on 6 February 2002.

The first reading opinion of the Parliament did not deviate too radically from the original proposals of the Commission. It did, however, already indicate some hesitation with regard to proposals on the renovations of existing buildings. The Parliament underlined the need for financial mechanisms to ensure energy efficiency measures could be implemented in renovations of existing buildings (European Parliament, 2002a: 2). The cost-effectiveness of the proposed measures, and the role of member states in incentivising the uptake of energy efficiency measures were emphasised in the opinion: thus the issue of financing was one of the main points for the Parliament in the ensuing negotiations. The Parliament also proposed new articles covering evaluation and information measures, to ensure follow-up of member state implementation and consumer awareness (ibid.: 11–12).

The Commission accepted several of the 30 amendments proposed by Parliament and published an amended proposal on 16 April 2002, including the emphasis of the Parliament on cost-effectiveness and some of the Parliament's suggestions for member state support of energy efficiency measures. These included tax deductions, low interest credit and funding through public programmes (European Commission, 2002a: 7–8). The Council position on this amended proposal followed on 7 June that same year.

Warnings had come earlier in the policy process that the Council was less interested in stringent measures on the energy performance of buildings than Parliament. Reports from an informal Council agreement in December 2001 suggested Council wanted to delay the implementation timetable of the measures in the proposed Directive by up to five years beyond the suggestion of the Commission. It also aimed to water down the requirements for existing building renovations, limiting the application of energy efficiency measures to 'major' renovations only (ENDS Europe, 2001b). It was therefore no surprise that the Council's position, published on 7 June 2002, included further flexibility for member states in applying the minimum standards for energy performance and added the proviso that standards will apply in existing buildings only undergoing 'major renovation' – a phrase left to the interpretive discretion of the member states themselves. The time period for implementing the provisions of the Directive was finally set by Council at 36 months, with additional flexibility so that member states could benefit from up to four additional years before fully implementing the provisions of the Directive (Council of the European Union, 2002: 13–14).

Not being able to agree on the deadline for the implementation of the proposed Directive, the Parliament began its second reading of the draft. The committee second reading recommendation was tabled for plenary on 11 September 2002, and the final second reading opinion of the Parliament was adopted on 10 October. News reports declared a 'soft' attitude of the Parliament towards the Council's amendments, voting in committee in favour of a six-year timetable for the implementation of the provisions of the Directive (ENDS Europe, 2002), rather than the quicker implementation it called for in its first reading (European Parliament, 2002b). The Commission and Council adopted the amendments proposed by Parliament in its second reading and the final text was signed on 16 December 2002.

Since then, there had been much criticism of the Directive, particularly in terms of the slow or insufficient implementation. In fact, commentators describe the Directive as having 'little impact' on the energy performance of buildings overall in the EU, due in part to its limited scope and in part to the great flexibility left to member states to implement the Directive as and when they wished (Henningsen, 2011: 133). Some stakeholders, including the insulation and glass industry associations, and environmental NGOs, were already calling for further legislation on the issue in 2004, citing insufficient action, poor implementation and difficulties to capitalise on the potential to reduce GHG emissions through the policy measures under the energy performance of buildings (ENDS Europe, 2004, 2005). These calls resulted in a proposal for a recast of the Directive.

Energy Performance of Buildings Recast Directive 2010/31/EU

The Commission published a proposal for a recast of the EPBD in 2008 (European Commission, 2008b). The proposal came in the context of discussions to move to a broader energy efficiency target to combat climate change, namely improving the energy efficiency of the EU by 20 per cent by 2020 (compared to business as usual projections) (European Council, 2007), but the EPBD recast proposal was not a part of the climate and energy package proposed in January 2008. The Commission's recast proposal was separate from the package, but negotiations on the EPBD recast and the package took place alongside each other (Boasson and Wettestad, 2013; Oberthür and Pallemaerts, 2010).

The recast Directive was considered necessary to 'extend the scope, simplify its implementation and develop energy performance of buildings certificates into a real market instrument' (European Commission, 2008c: 11). The proposal for a recast of the EPBD, which was released together with the second strategic energy review, kept the general framework of the first EPBD focusing on calculation methodologies; recommendations for financial support; certification measures; requirements for heating and cooling systems; and requirements for national building codes. It did also extend the 2002 Directive by proposing that *all* buildings undergoing major renovations would have to meet minimum energy performance standards (no longer large buildings); that energy certificates would have to be used in all sales and rental advertisements; and that the inspection of boilers would have to be improved. The proposal also encouraged member states to develop national policies for the greater uptake of low- and zero energy buildings (European Commission, 2008b: 6).

The Parliament took up the first reading of the proposal, again with the ITRE committee drafting the first reading opinion. Silvia-Adriana Ţicău, a Romanian MEP from the group of the Progressive Alliance of Socialists and Democrats (S&D), acted as rapporteur. Among the five shadow rapporteurs, MEP Claude Turmes represented the Greens. The first reading opinion, adopted by the Parliament on 23 April 2009, was already more ambitious than the Commission proposal in many respects. The rapporteur pushed for financial incentives to encourage energy efficiency measures, and MEP Fiona Hall, member of the Alliance of Liberals and Democrats for Europe political group (ALDE) and shadow rapporteur, pushed for tightening proposals to include targets for all new buildings to be zero energy by 2020 (Boasson and Wettestad, 2013; ENDS Europe, 2009d). The final first reading report, with 107 amendments to the Commission's proposal, included both of these proposals, with 2019 as the suggestion deadline for zero-energy buildings and a call for an energy efficiency fund to be created by 2014 (ENDS Europe, 2009e; European Parliament, 2009a, 2009b).

Member states in the Council were less enthusiastic about strengthening the Commission's proposal for a recast, and there were early indications of reticence in applying minimum standards to all buildings undergoing major renovations – a major element of the Commission's proposal (Boasson and Wettestad, 2013; ENDS Europe, 2009c). The debate in Council took place in its transport, telecommunications and energy formation on both 12 June and 7 December 2009. The Council published its common position on 14 April 2010 (Council of the European Union, 2010). In the run-up to the adoption of its agreed position, the Council complained that Parliament's amendments were overly ambitious

(ENDS Europe, 2009b). Informal negotiations with the Parliament (in trialogues, with the rapporteur of the ITRE committee) were ongoing until the adoption of its position (Council of the European Union, 2010: 55–56; ENDS Europe, 2009f). The final position showed the fruits of these negotiations: the deadline for achieving 'nearly-zero-energy buildings' was set at 2020, with public buildings to achieve this goal in 2018. 'Nearly-zero-energy buildings' refers to achieving a very highly-efficient energy performance, the result of a compromise between Parliament and Council. However, it was up to the member states to decide on the standard meant by 'nearly-zero-energy'.

The Parliament responded very quickly with its second reading recommendation, published on 18 May 2010, approving the Council's position, leading to the adoption of Directive 2010/31/EU.

CPI into the policy process of the 2002 EPBD (2002/91/EC)

As with the discussion on CPI into the EU's renewable energy policy (see Chapter 3), I measure the integration of climate policy objectives into the policy process by analysing the involvement of internal and external pro-climate stakeholders and the level of recognition of functional interrelations between buildings policies and climate policy objectives.

Internal pro-climate stakeholders

Internal pro-climate stakeholders in the policy process leading to the 2002 EPBD only had a *low* level of involvement, but the actors who were involved expressed conviction about the need to agree policies to improve the energy performance of buildings to combat climate change. In the Commission, DG Energy drafted the proposal for the 2002 EPBD. Internal procedures within the Commission allowed DG Environment to comment on the proposal. When the proposal came to the college of commissioners, three commissioners objected to the need for a Directive on the issue based on subsidiarity grounds: Chris Patten, then commissioner for external relations, Neil Kinnock, then Vice President of the Commission, and Frits Bolkestein, then commissioner for the internal market and services. Although they were overruled, their objection led to further time redrafting the proposal before its publication. The redrafting was again in the lead hands of DG Energy, under Commissioner Loyola de Palacio. DG Environment did not play a particularly active role in pushing for more ambitious measures than were already outlined by DG Energy. This implies a rather *low* level of involvement of internal pro-climate stakeholders in the Commission, with DG Environment able to provide opinions to the process if it wished, but not playing a major role in the internal consultations in the Commission. This low level is tempered by the climate motivations of DG Energy, yet these motivations were nonetheless compromised with the objections of other commissioners.

In the Parliament, although the ENVI Committee provided an opinion to the ITRE committee, it did not seem to influence the final weakened Directive. The main amendments in the Parliament related to funding, cost-effectiveness and implementation deadlines. Although the ENVI Committee had the procedural option of providing an opinion, this did not lead to substantive influence to strengthen the proposal. The ENVI Committee suggested amendments that were not taken on board. One of these proposed amendments specified that buildings with a floor area of 500m² undergoing renovations would be required to meet minimum energy performance standards (European Parliament, 2001: 31). This particular proposal could have had a substantial impact on the ability of the EPBD to achieve its expected policy output. The final decision to set the threshold for energy efficient renovations for buildings with a floor area of more than 1000m² considerably reduced the scope of application of the Directive. Thus, the involvement of pro-climate stakeholders in the Parliament is *low to medium*, where the ENVI Committee provided an opinion, but this opinion was not necessarily taken on board or reflected in the process.

In the Council, the discussions on the legislative proposal took place within the Council of energy ministers. The proposal was first discussed in the industry and energy Council meeting in December 2001, where the general outlines of the Council's position to promote further member state flexibility in the final Directive were already agreed (Council of the European Union, 2001; ENDS Europe, 2001b). The next occasion when the issue was discussed led to the adoption of the common position, again in the industry and energy Council meeting in May 2002 (Council of the European Union, 2002). The Council approval of the final text occurred in the transport, telecommunications and energy council meeting in November 2002. As is clear throughout the process, the environment Council was not involved in the development of the Council's position. In fact, member states' concerns on this file were linked to financial implications and issues of subsidiarity. The climate policy motivations for the policy rarely entered discussions in the Council. This leads to a *very low* level of involvement of internal pro-climate stakeholders in the policy process in the Council.

Overall, and in aggregate, a *low* level of involvement of internal pro-climate stakeholders in the EPBD policy process is evident (see Table 2.1). While some opportunities existed for pro-climate stakeholders to be involved in some of the institutions, either they could not or did not take advantage of these opportunities or their opinions were disregarded in the policy process. In some respects, as combating climate change was one of the stated motivations behind the EPBD proposal (along with strong statements on improving energy security), some level of policy integration for the main actors in the policy process could have been assumed. However, as is clear from the poor final output and performance of the EPBD (see above), the low level of involvement of climate stakeholders may not have helped climate policy advance through buildings policy.

External pro-climate stakeholders

The level of involvement of external pro-climate stakeholders in the policy process leading to the 2002 EPBD is *low*. First, it is clear that environmental

NGOs working at the EU level in the early 2000s did not focus on energy efficiency measures. In fact, interviewees have indicated that none of the major environmental NGOs had staff dedicated to working on energy efficiency at this time (interviews 5, 6, 10 and 11). Therefore, environmental NGOs responded to developments as they occurred rather than spurred action at EU level. The task fell to already overburdened staff within the NGOs to follow the developments on this dossier (interviews 5 and 6). These staff members were able to respond only in a limited way and only when major developments occurred in the institutions. Greenpeace, for example, came out in support of the Parliament in its first reading, when it seemed the Parliament would resist Council pressure to extend the implementation deadline of the Directive (ENDS Europe, 2001a). Opportunities did exist for informal lobbying during the policymaking process, but environmental NGOs rather lacked the expertise and the resources to push policymakers to make strong pro-climate policy.

Second, industry associations in favour of energy efficiency measures for buildings were only in their early development at this time. Three industry associations were active and provided some input to policymakers. These included EuroACE (the European Alliance of Companies for Energy Efficiency in Buildings), FIEC (the European Construction Industry Federation), and Eurima (the European Insulation Manufacturers Association). Each of these associations lobbied the EU institutions to enact strong legislation on the energy performance of buildings, underlining job-creation opportunities and climate benefits (ENDS Europe, 2001b). Nevertheless, FIEC, EuroACE and Eurima were more 'reactive' than 'proactive' in the policy process leading to the 2002 EPBD (Boasson and Dupont, 2015; Boasson and Wettestad, 2013: 135; interview 10). One interviewee described how his organisation lacked effectiveness at lobbying the EU institutions in the early 2000s, saying they 'never quite dotted the i's and crossed the t's on that first Directive' (interview 10). Although these associations reacted against warnings that the Council would extend the implementation timetable, they could not influence the final output. Many of these associations would go on to call for a tightening of the legislation on energy performance of buildings not long after it was agreed (ENDS Europe, 2004, 2005).

Third, there were few official procedures and avenues for stakeholders to access policymakers beyond their own informal lobbying activities at this time. The legislative proposal was made in 2001, before regulatory impact assessments and public consultations were introduced in the Commission in 2002 (European Commission, 2002b). The Commission's proposal did not indicate any formal or informal consultation with external actors. It is clear that some points in the Commission's proposal came from working groups of the early European Climate Change Programme (ECCP) process, which aimed to involve member state, industry and NGO representatives in discussions on potential policy measures to combat climate change (European Commission, 2000, 2001). This working group process was rather opaque. It is unclear which organisations were involved in pushing policy in the energy performance of buildings particularly. However, industry members and member states (who may not necessarily promote climate policy objectives) certainly had an extra opportunity to express their interests in the working group on sustainable construction set up in 2001, where environmental NGOs were not present (Boasson and Wettestad, 2013).

The level of external climate policy stakeholder involvement in the policy process can thus be considered *low* (see Table 2.1). Pro-climate stakeholders, whether from environmental NGOs or from industries interested in pushing for ambitious policy, did not play much of a role in the policy process and had little influence. The peculiarity of this time period is the lack of dedicated expertise or resources among environmental NGOs on energy efficiency issues, meaning that it was the industry associations working on energy efficiency measures that played a more substantial (yet still limited) role in the process. These industry associations were more likely to use job-creation arguments in their lobbying than climate policy arguments. However, even the industrial actors did not play a major role in the policy process in the lead up to the 2002 EPBD (interviews 5, 6, 10, 11 and 12).

Recognition of functional interrelations

There is a *medium* level of recognition among policymakers of functional interrelations between climate policy objectives and the 2002 EPBD objectives. From the very outset of the policy process, policy on the energy performance of buildings was linked to combating climate change and achieving the objective of reducing GHG emissions in line with the EU's commitment under the Kyoto Protocol. In its proposal, the Commission mentioned that concerns about climate change, and concerns about the rising dependence of the EU on energy imports, were reasons for economising the 'use of energy wherever possible' (European Commission, 2001: 2). Recital 3 of the final Directive also demonstrates the general understanding by the Commission, Council and Parliament that the EPBD responded to the challenge of climate change: 'increased energy efficiency constitutes an important part of the package of policies and measures needed to comply with the Kyoto Protocol' (Directive 2002/91/EC). The functional interrelations between reducing energy consumption in buildings and combating climate change by reducing energy-related GHG emissions were clearly recognised and were highlighted, at least at the outset of the process.

Compared to the long-term climate policy objectives of reducing GHG emissions in the EU by 80 to 95 per cent by 2050 or ensuring that global temperature increase is limited to 2° Celsius compared to pre-industrial levels, however, neither the proposal drafted by the Commission, nor the amendments by Parliament and Council, nor the final Directive itself can be considered ambitious enough. The Directive does not sufficiently emphasise improvements in the existing building stock, which is crucial for the achievement of significant reductions in time for 2050. Indeed, the EU's building stock includes 40 per cent of buildings that were built before 1960, and many buildings that were built in 1910 or earlier (BPIE, 2011), when energy performance standards were non-existent. Additionally, the Directive outlined that it was up to member states to decide on and implement the energy performance standards for new buildings (Recitals 9, 10, 12; Arts. 3, 4). The EU institutions did not translate the recognition of the functional interrelations between the energy performance of buildings policy and climate policy objectives into ambitious policy measures. Neither did they consider long-term climate policy measures beyond the Kyoto Protocol commitments (to reduce GHG emissions in the EU15 by 8 per cent by 2008–2012 compared to 1990 levels). In line with the indicators outlined in Table 2.1, a *medium* level of recognition of the functional interrelations is assigned in this case. Functional interrelations were recognised at the beginning of the policy process, but this recognition did not motivate more ambitious policy measures.

In sum, the level of CPI in the policy process leading to the adoption of the 2002 EPBD is *low to medium*, with *low* involvement of climate stakeholders but *medium* recognition of functional interrelations.

CPI in the policy output of the 2002 EPBD

Energy consumption in buildings stood at 447 Mtoe in 2000. With energy consumption of the EU's buildings rising from 423 Mtoe in 1990 to 447 Mtoe in 2000, continuing such a trend would lead us to expect energy consumption of buildings to hit about 471 Mtoe in 2010 without any policy measures, under a BAU scenario. In its proposal for the EPBD, the Commission stated that the potential for improvement in the energy performance of buildings could amount to an energy consumption reduction of 22 per cent between 2000 and 2010 (European Commission, 2001: 8). From the 447 Mtoe of energy consumption in buildings in 2000, a 22 per cent reduction is equal to a reduction of about 98 Mtoe by 2010 – meaning energy consumption in buildings could achieve 349 Mtoe in 2010 with strong enough EPB policy. The Commission regarded this as a feasible, but ambitious, reductions potential. The Commission goes on to state that the indicative energy savings target agreed in the Council's Resolution on 7 December 1998 on energy efficiency (98/C 394/01) to improve the energy intensity of final consumption in the EU would result in avoided energy consumption of about 55 Mtoe by 2010 (in other words, avoided consumption compared to what would otherwise have been consumed). Avoiding 55 Mtoe of BAU expected levels of 471 Mtoe energy consumption in buildings in 2010, through measures of the EPBD, would result in 416 Mtoe energy consumption of buildings in 2010. This was considered the real expected output of the EPBD. Thus, the 2002 EPBD was expected to reduce energy consumption of buildings to at best 416 Mtoe by 2010 (see Figure 4.4).

A policy output that avoids 55 Mtoe of energy consumption in buildings by 2010, compared to BAU, would succeed in closing part of the gap between expected BAU and CPI scenarios to 2010. Energy consumption of buildings was expected to hit 471 Mtoe in 2010 without the EPBD. The range of scenarios for CPI for 2010 is between 377 and 397 Mtoe of energy consumption in buildings by 2010 (see also Figure 4.3). Thus, the policy output in the 2002 EPBD (of 416 Mtoe in 2010) closes the gap between BAU (471 Mtoe) and CPI ranges by

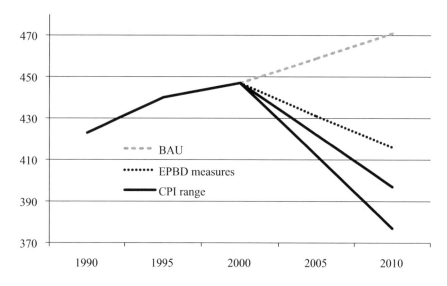

Figure 4.4 CPI in the policy output of the 2002 EPBD, with measures expected to avoid the consumption of 55 Mtoe of energy in buildings in the EU between 2000 and 2010

Source: Compiled from (European Commission, 2001), own calculations

between 59 and 74 per cent. As outlined in Table 2.2, this results in *medium to high* levels of CPI for the policy output of the 2002 EPBD.

This cannot be considered a realistic measurement for the level of CPI, however. First, implementation of the EPBD provisions was delayed in member states by between three and seven years after its entry into force. Second, the expected impact of the EPBD is based on a non-binding indicative target. Later analysis shows the inadequate implementation across the EU of the EPBD measures, and much literature describes the EU's implementation deficit (see Boasson and Dupont, 2015; Haverland and Romeijn, 2007; Lampinen and Uusikylä, 1998). Third, actual energy consumption of buildings between 2000 and 2010 *rose* significantly (from 447 Mtoe in 2000 to 482 Mtoe in 2010), even beyond the BAU scenarios suggested in Figure 4.4. It is difficult to see how the EPBD had any positive effect on the reduction of energy consumption, considering this reality. At best, it could be argued that the EPBD provided a first step towards later policy improvements, and may have helped ensure the increase in energy consumption was not even higher (see Figure 4.5).

A very high CPI trajectory range shows the need for energy consumption to be reduced between 2000 and 2010 by between 50 and 70 Mtoe, but actual energy consumption in the same time period increased by 36 Mtoe (Figures 4.3, 4.4 and 4.5). Energy consumption of buildings did fall by (just) 2 Mtoe between

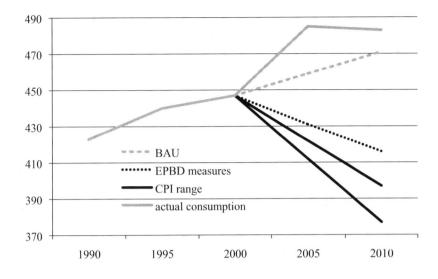

Figure 4.5 EPBD measures compared to actual consumption of energy in buildings, 2000–2010 (measured in Mtoe)

Source: Compiled from (European Commission, 2001; Eurostat, 2015), own calculations

2005 and 2010 (European Commission, 2011c). Energy consumption had risen between 1990 and 2005 by 62 Mtoe, so one could assume that without the implementation of the EPBD, energy consumption might have continued to rise along the same trend to 2010 and beyond. Thus, we could have expected to see on average a further 21 Mtoe of energy consumption in the buildings sector between 2005 and 2010, which would have resulted in energy demand in the buildings sector of 506 Mtoe by 2010. Instead, the actual energy consumption of buildings was recorded at 23 Mtoe lower in 2010 (483 Mtoe). However, while some of this reduction could potentially be attributed to a functioning EPBD, much of this avoided energy consumption can also be considered a direct effect of the financial and economic crises from 2008 onwards, placing doubt on whether or not the EPBD had even this limited impact in this time period (Henningsen, 2011). While a reduction of 2 Mtoe between 2005 and 2010 (to 483 Mtoe) could be considered in a positive light, it remains difficult to assess the extent to which the EPBD played a role. Even with the assumption that the EPBD was a major factor in the small reduction in energy consumption, the level in 2010 is still a considerable way from a high CPI level of between 377 and 397 Mtoe. In fact, if BAU for 2010 hit 506 Mtoe, and the actual policy impact of the EPBD was the avoidance of 23 Mtoe, then the EPBD could be considered to have closed the gap between BAU and CPI ranges for energy consumption of buildings by about 18 to 22 per cent. Even taking the EPBD output in a positive light, such a measurement would arrive at CPI in the policy outcome/impact of the EPBD of *very low to low* levels.

In conclusion, a reasonable assessment of the level of CPI in the policy output for the 2002 EPBD would lie somewhere between the expected results (*medium to high levels* of CPI) and the empirical reality of the EPBD's impact (*very low to low* levels of CPI). The expected output was tempered by delayed and poor implementation and due to the non-binding nature of the energy efficiency target. A generous assessment of CPI in the policy output implies taking account of these tempering factors, while also giving the benefit of the doubt to the ambitions for the policy output of the EU in agreeing the Directive in the first place. However, it is impossible to ignore the time-delay for implementation of the EPBD provisions and the non-binding nature of the final policy output that made any achievement of the potential of the EPBD doubtful. Such a consideration places slightly more weight on the final *very low to low* levels of CPI. Thus, the level of CPI in the policy output of the 2002 EPBD could be described as *low*.

CPI into the policy process of the 2010 EPBD recast (2010/31/EU)

Internal pro-climate stakeholders

The involvement of internal pro-climate stakeholders in the policy process leading to the adoption of the 2010 EPBD recast is *low*. It was again DG Energy in the Commission, the ITRE committee in the Parliament, and the transport, telecommunications and energy Council that led the negotiations. However, through the usual consultation procedures, there were opportunities for internal pro-climate stakeholders to at least be involved in the policy process. Additionally, the main actors could be considered as taking climate policy objectives on board, since they described combating climate change as part of the motivation for improving policy on the energy performance of buildings.

In the Commission, DG Environment was one of eight DGs involved in the impact assessment process. Many Commission DGs were in favour of more ambitious policy on the energy performance of buildings at this time. It is plausible that DG Environment did not see the need to be greatly involved, especially with DG Energy pushing for ambitious policy (interview 12). This implies a *very low to low* level of involvement of internal pro-climate stakeholders in the Commission, as, considering the procedures allowed for consultation, the pro-climate actors were not visible in the process. But these actors themselves did not seem to take advantage of such opportunities as they may not have seen the need to do so. The *very low to low* level may thus be too harsh a judgment considering that pro-climate voices internally were present also in DG Energy.

Within the Parliament, the evidence points to more interaction with pro-climate stakeholders, although not in terms of ENVI Committee involvement. The rapporteur on the dossier had a pro-climate stance, and became a member of the cross-party renewable energy association of parliamentarians, EUFORES.

In addition, shadow rapporteur and Green MEP Claude Turmes was often outspoken on the need to improve the environmental ambition of the proposal and to promote energy efficiency as a top priority (ENDS Europe, 2009a, 2010c). MEP Fiona Hall, of ALDE, was also influential in the policy negotiations in pushing for more ambitious policy (Boasson and Wettestad, 2013). In this case, the level of involvement of pro-climate stakeholders in the Parliament is *low to medium*, with pro-climate shadow rapporteurs playing a role in the drafting of the Parliament's reports and opinions, but the ENVI Committee being rather absent. As with the Commission, the absence of explicit pro-climate actors may not necessarily reflect the reality of the level of CPI in the policy process. In the Parliament, those actors that were involved were generally in favour of stronger policy on the energy performance of buildings, with combating climate change being one of the main aims.

Within the Council, the main pro-climate stakeholder involved in the policymaking process can perhaps be identified as the Presidency at the time of the final negotiations – Sweden (in the second half of 2009). The environmental Council was not involved, but the general green credentials of Sweden (Liefferink, Arts, Kamstra and Ooijevaar, 2009) might have helped push the dossier through. Although the official agreement on the dossier did not come until the publication of the Council's common position in April 2010 and the Parliament's adoption of this position in second reading, the negotiations in trialogues in the late part of 2009 meant that an informal agreement had been reached already in December 2009 – in time for the important round of international climate negotiations in Copenhagen. This was considered a success for the Swedish Presidency (interview 6). However, there was no real visible role of the environment Council in the negotiations, and member states were generally the most reticent actors in agreeing on strong policy on the energy performance of buildings. The special facilitating role of the Presidency means that the level of pro-climate stakeholder involvement in the Council at this time is estimated as *low* (rather than *very low*).

As much of the final negotiations took place behind closed doors in trialogues, among the Presidency, the rapporteur and shadow rapporteurs from the Parliament and the Commission, it is difficult to assess accurately the interaction of pro-climate stakeholders throughout the policy process. One interviewee emphasised that the Parliamentary representatives are often at a disadvantage in trialogues, due to lacking negotiation skills, resources and expertise on the topic, among others (interview 5). On the EPBD recast dossier, the Parliament may have held the most ambitious environmental stance, but in the name of agreement and compromise, many of the Parliament's green voices were left outside during the trialogue negotiations.

Overall, in the case of the involvement of internal pro-climate stakeholders in the policy process leading to the agreement on the recast of the EPBD, a *low* level of involvement can be ascribed. None of the traditional pro-climate stakeholders were in the lead in policy negotiations, although some of the pro-climate arguments were taken over by other actors perhaps rendering it less necessary for pro-climate stakeholders to be very involved. Regular procedures were in place to allow access to the policy process in the Commission and in the Parliament, but access was then limited in the Council and in the trialogue negotiations. Internal pro-climate stakeholders' arguments were represented by some of the main policymakers, but the ENVI Committee did not give an opinion on the dossier, and neither was it discussed in the environment Council. While internal pro-climate stakeholders might have been involved in early phases of the policy negotiations, the later negotiations in trialogues excluded all actors outside those directly negotiating.

External pro-climate stakeholders

The involvement of external pro-climate stakeholders in the policy process leading to the 2010 EPBD recast is *medium*.

By the late 2000s, environmental NGOs had started to pay attention to the energy efficiency issue and to invest in influencing the policy process. Interviewees from the environmental NGO community in Brussels freely admitted that the lack of resources was a barrier that previously played against them. However, although energy efficiency was an issue that was taking more of their attention, they were in a mode of learning and still suffered from some resource constraints for the recast of the EBPD (interviews 5 and 6). Two interviewees also complained about difficulties in knowing the timetable of work of the Commission to be able to provide direct input at the right moment (interviews 5 and 11).

In the run-up to the publication of the Commission's proposal, several conferences and formal consultations with stakeholders took place (European Commission, 2008a: 15–17). These consultation procedures allowed access for external pro-climate stakeholders to the drafting of the proposal stage in the policy process. An online consultation took place for eight weeks from 25 April 2008, and received 246 responses, with organisations and citizens from 22 member states responding. Fifty of these responses came from NGOs, another 33 from associations and 83 responses came from industry. Institutions and member states provided 24 responses, 44 responses came from citizens and the remaining five responses were filed under 'other' (European Commission, 2008a: 5, 16, annex 1). The Commission also highlighted other occasions where it interacted with member states and industry representatives to discuss options for revising the 2002 EPBD, such as concerted actions under the Intelligent Energy Europe programme in 2006–2007, and during a conference organised as part of the EU sustainable energy week in 2008. Thus, consultation procedures were in place and were used by environmental NGOs and pro-climate industrial stakeholders (and many other external stakeholders) leading up to the proposal of the Commission on the recast of the EPBD. Environmental stakeholders were not necessarily in the majority, but they certainly had opportunities to be present in the run-up to the Commission's publication of the proposal.

Many NGO representatives rather chose to invest their resources and effort into influencing the opinions of the Parliament, however, which interviewees considered as a more open institution (interviews 5, 6, 7 and 8). As one interviewee said, well-resourced environmental NGOs 'can have an impact on the Commission ... but that's only before the proposal goes through the inter-service consultation ... where it can then be changed completely', whereas all NGOs 'can access MEPs quite easily' (interview 5). This ease of access to the Parliament led to what this particular interviewee described as the 'best first reading ever' from an environmental NGO perspective, in terms of the improvements suggested through the amendments proposed by the Parliament.

Beyond the Parliament, environmental NGOs described great difficulty in influencing negotiations in the Council, and especially the trialogue negotiations that took place in the final months before the adoption of the Council position. 'These meetings [trialogues] are done behind closed doors', complained one interviewee, 'there's a lot of lost transparency and it's difficult to keep track of what's going on' (interview 6). Environmental NGOs had little access to policymaking in the Council and to the final stages of the negotiating process where no procedures exist to ensure access to the policy process.

With regard to pro-climate stakeholders from industry (in this case, industries interested in pushing for ambitious energy efficiency measures), those industry associations involved in the first EPBD were also present in the negotiations leading to the EPBD recast. Pro-climate industry stakeholders were more organised at the EU level in the negotiations on the EPBD recast (Boasson and Wettestad, 2013: 142), and even linked with environmental NGOs in joint campaigns. The coalition for energy savings, for example, brings together 26 industrial associations and environmental NGOs to push for more ambitious policy measures on energy efficiency. It was hoped that the coalition would have a clear impact on the policy process, with industry and environmental voices presenting the same message (for more ambitious policy). However, the nature of the buildings industry (dominated by local and small enterprises) meant that the message put forward by the industry associations in Brussels was only part of the buildings' industry's voice (Boasson and Dupont, 2015).

While environmental NGOs found better access to the Parliament, interviewees from industrial associations stated a clear preference for lobbying the Commission (interviews 10 and 11). The Commission, as one interviewee put it, is interested in 'fact-based representation from industry', while the Parliament is far more political, making it more difficult for industry representatives to position themselves in lobbying the Parliament (interview 11). Working together with environmental NGOs meant that a common message was delivered to both the Commission and the Parliament. Interestingly, stakeholders clearly mention that at the time of the negotiations leading to the recast EPBD, there were 'no major stakeholders opposing the EPBD', but that it was rather reticence from member states that posed the main barrier to agreeing stronger legislation (interviews 5, 6 and 10). Stakeholders were at a disadvantage in terms of trying to influence the really reticent actors, considering the lack of access to the trialogue negotiations. While industry and environmental stakeholders were positive about the improvements made in the recast of the EPBD, they nevertheless described dissatisfaction with the final output, and the insufficiency of the new measures to meet the (non-binding) 20 per cent energy efficiency target by 2020 (ENDS Europe, 2010a, 2010b).

Overall, the level of involvement of external pro-climate stakeholders in the policy process is *medium*. Procedures existed in the preparation of the proposal that allowed external stakeholder involvement (public and online consultations). Such procedures did not exist in later stages of the process, and stakeholders complained of lack of access to the Council and to the policymaking negotiations taking place in trialogues. The pro-climate lobby faced opposition from policymakers in the Council especially. The Parliament's first reading opinion was warmly welcomed by pro-climate stakeholders, but it was later substantially watered down during the trialogue negotiations. The internal Council negotiations and the trialogues turned out to be venues for serious negotiations on this policy process, and external pro-climate stakeholders had little access to, or involvement in, these venues.

Recognition of functional interrelations

The recognition of the functional interrelations between climate policy objectives and EPB policy objectives in the policy process leading to the 2010 EPBD recast is *medium*. Both the Commission and Parliament can be regarded as recognising the interrelations between EPB policy and climate policy objectives and pushing for stronger EPB policy as a result (especially in the Parliament). While the Council may have rhetorically recognised the interrelations between climate policy and energy performance of buildings policy, this recognition did not lead to the Council backing strong policy measures. In this respect, we can see a difference in the recognition of short-term policy objectives and long-term objectives to achieve 80–95 per cent GHG emission reduction by 2050. In all three institutions, the long-term policy objectives were less a part of the negotiations than shorter-term aims (to 2020, for example).

From the outset, the Commission highlighted the environmental and climate benefits of improving the energy performance of buildings in the EU (European Commission, 2008b). The recast did not change the objectives of the EPBD, but rather aimed to clarify further concepts and tighten measures so that the EPBD could do more than simply bring energy efficiency onto political agendas, which is what the Commission considers to have been the main contribution of the 2002 EPBD (ibid.: 3). Thus, the Commission follows the 2002 wording in the proposal for the EBPD recast on its potential to contribute to meeting the EU's commitments under the Kyoto Protocol and to combat climate change generally, and also as part of the contribution to the non-binding energy savings target to 2020, but there is no explicit mention of long-term climate policy objectives.

In Parliament, there was a clear interest to push for ambitious policy on the energy performance of buildings. The first reading agreement proposed amendments to the Commission's proposal that would have included wording about the EU's long-term climate commitment to limit global temperature to 2° Celsius in the opening recitals of the text. The Parliament also called for the 20 per cent energy efficiency target to 2020 to become binding; called for increasing the number of zero-energy buildings in the EU; and proposed measures to increase financing for improving energy efficiency.

The Council did not accept all these ambitious amendments, and the final EPBD outlined that new buildings would be 'nearly-zero energy' by 2020 and public buildings by 2018 – with the definition of 'nearly-zero' left to the discretion of the member states. Member states weakened Parliament's proposals for financing and would not accept any binding efficiency targets (see also Boasson and Wettestad, 2013). The Parliament's amendment to include the long-term commitment to limiting global temperature increase to 2° Celsius remained throughout the negotiations (Recital 3, Directive 2010/31/EU). Yet this recognition of long-term functional interrelations was not backed up with the sort of ambitious policy measures that would then be required. In other words, the inclusion of Recital 3 did not translate into ambitious energy efficiency measures to achieve 80 to 95 per cent reductions in GHG emissions by 2050, compared to 1990 levels.

In line with Table 2.1, the recognition of functional interrelations in the case of the 2010 EPBD recast is *medium*. While both short-term and long-term climate policy objectives were considered throughout the policy process, and are part of the stated motivations of the policy measure, the recognition of these functional interrelations did not lead to or motivate policy in an ambitious direction. While Council may be most at fault in this case, it may be that Parliament gave in to Council demands too easily without considering the ramifications for long-term climate policy.

In summary, taking account of the three measurements outlined above, the level of CPI in the policy process is measured, overall and on aggregate, as *low to medium*.

CPI in the policy output of the 2010 recast EPBD

In its proposal, the Commission outlined that the potential for measures under the recast of the EPBD could amount to the avoidance of between 60 and 80 Mtoe of energy consumption by 2020, compared to BAU (European Commission, 2008b: 4). With the disappointing results of the 2002 EPBD, the BAU scenarios had changed by this time. According to the 2006 action plan for energy efficiency, BAU scenarios foresaw 549 Mtoe energy consumption in buildings in 2020 (European Commission, 2006: 6). The EPBD recast proposal could then be expected to reduce this level by between 60 and 80 Mtoe, to between 469 and 489 Mtoe. As discussed above, very high CPI means achieving a maximum of 200 Mtoe of energy consumption in buildings by 2050. With 2010 consumption standing at 483 Mtoe, this implies a steep reduction in energy consumption. Under very high CPI scenarios, therefore, energy consumption of buildings in 2020, starting from actual consumption rates in 2010, should be no more than 387 to 412 Mtoe. Thus, the proposal for the recast of the EPBD closes the gap of 137 Mtoe between BAU at 549 Mtoe and CPI of 412 Mtoe by between 44 and 58 per cent. It closes the gap of 162 Mtoe between BAU at 549 Mtoe in 2020 and CPI of 387 Mtoe by between 37 and 49 per cent. This is represented in Figure 4.6. According to Table 2.2, this range of gap-closure of 37 to 58 per cent represents a *low to medium* level of CPI in the policy output of the EPBD recast.

As discussed above, other factors may temper the reality of achieving these objectives to 2020. These factors are similar to those that resulted in the very low impact of the 2002 EPBD, such as the lack of a binding target and poor implementation by member states. In the case of the 2010 EPBD recast, there was still no binding target for achieving energy efficiency measures, and fears were already being circulated at the agreement of the EPBD recast that it was unlikely the overarching 20 per cent energy savings by 2020 target would be achieved (Boasson and Dupont, 2015; Henningsen, 2011). The Commission later revised its estimates on the impact of the EPBD in its energy roadmap to 2050 (European Commission, 2011c). Even with the EPBD recast in force, the energy consumption of buildings is still expected to rise by 2015 to nearly 500 Mtoe in the Commission's 2011 scenarios, before falling only incrementally to approximately 496 Mtoe in 2020 (and only 472 Mtoe in 2050; see Figure 4.7).[1]

With 2006 BAU scenarios suggesting that energy consumption of buildings would reach 549 Mtoe, the new BAU scenario from 2011 that includes the

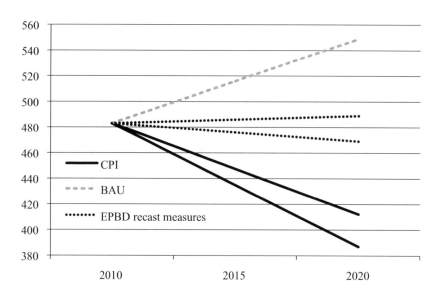

Figure 4.6 CPI in the policy output of the 2010 EPBD recast, with measures expected to avoid between 60 and 80 Mtoe of energy consumption in buildings in 2020 (compared to BAU and very high CPI range; measured in Mtoe)

Source: Compiled from (European Commission, 2006, 2008), own calculations

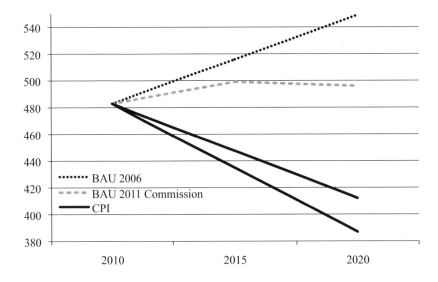

Figure 4.7 Updated BAU scenario from 2011, which includes expected EPBD impact on the energy consumption in buildings to 2020, compared to the 2006 BAU trajectory and CPI ranges to 2020 (measured in Mtoe)

Source: Compiled from (European Commission, 2006, 2011; Heaps et al., 2009), own calculations

impact of EPBD suggests the EPBD recast would avoid the consumption of 53 Mtoe in 2020. This closes the gap between complete CPI trajectories by between 33 and 38 per cent. Thus CPI in the policy output could be measured as *low*, according to Table 2.2. Considering the poor implementation of the 2002 EPBD, a *low* level may prove a best case for the EPBD recast if member states follow a similar pattern of inadequate implementation. Overall, then, with best estimates of *low to medium* and more updated estimates of *low* CPI, the *low* score may be more justified.

Summary and outlook to 2030 and beyond

EU policy on the energy performance of buildings is particularly interesting for assessing the level of CPI in the EU's energy sector. The EPBD and its recast were promoted as energy policy measures achieving a multitude of objectives, including improving energy security, combating climate change, and creating jobs. Since these measures have been in place, they have sometimes come to be regarded as climate policy measures, and it is indeed difficult to imagine a policy measure on energy efficiency that does not help meet climate policy objectives.

In this chapter, I examined CPI in the policy output and process of the 2002 EPBD and of the 2010 EPBD recast. These policies included measures to improve the energy performance of buildings by targeting the building envelope and the energy efficiency of installed equipment (such as hot-water boilers) and by requiring the energy certification of buildings. Both the 2002 and 2010 Directives focused on measures to improve the energy performance of buildings in new buildings, and in buildings undergoing major renovations only. The 2002 Directive included a 1000m^2 floor area threshold for building renovations that were obliged to meet minimum energy performance standards. In the 2010 Directive, this threshold was removed, and all buildings undergoing 'major renovations' are expected to meet these minimum standards. Both Directives leave implementation and interpretation up to the discretion of the member states. Definitions of minimum energy performance standards, of nearly-zero energy standards, and of what constitutes cost-optimal measures are not provided in the Directives, and member states have the flexibility to implement their own interpretations of these terms. Additionally, no binding energy efficiency targets were agreed in either the 2002 or 2010 Directives.

Although assumptions may exist that policy measures on energy efficiency can be considered climate policy measures, the analysis in this chapter clearly demonstrates the inadequate levels of CPI in the policy processes and outputs of the 2002 and 2010 EPBDs. With *low to medium* levels of CPI in the policy processes and *low* levels of CPI in the policy outputs of both Directives, these policies do not demonstrate sufficient levels of CPI to help achieve long-term climate policy objectives to 2050. The history of the development of the EPBD shows that combating climate change was a late motivation for policy development. Energy security concerns dominated energy efficiency policy objectives in the 1970s and 1980s, and this motivation remained part of the policy objectives after combating climate change was added. The 2002 EPBD was very much modelled on the policy measures outlined in the Commission's 1984 communication on the rational use of energy in buildings, where neither environmental nor climate objectives are considered (European Commission, 1984).

Since the adoption of the recast of the EPBD, the EU has agreed a further policy measure on energy efficiency – namely the 2012 Energy Efficiency Directive (EED; Directive 2012/27/EU). Under this Directive, member states are required to set 'indicative national energy efficiency targets' (Recital 13, Art. 3.1). Targets for energy efficiency thus remain non-binding. The EED does advance policy on buildings, most notably by requiring central government buildings to be renovated at a rate of 3 per cent per year (Art. 5) and by requiring renovations roadmaps for each state's entire building stock (Art. 4). However, despite the addition of the 2012 EED, the 2020 target to improve energy efficiency overall in the EU by 20 per cent is still expected to require further policy measures for achievement (Boasson and Dupont, 2015; EEA, 2014).

Energy efficiency is part of the proposed 2030 framework on climate and energy. In October 2014, the European Council agreed to a non-binding target of 'at least 27 per cent' energy efficiency improvements compared to 'projections of future energy consumption', which will be reviewed in 2020 with a 30 per cent target in mind (European Council, 2014: 5). Considering the high potential for energy efficiency in the EU, the long-term climate policy objectives to 2050 and the need for early action on climate change, and the past record on EU policy on buildings, a more ambitious policy target would be more appropriate. Furthermore, the proposed target remains non-binding, despite continued failures of the EU to meet its historical non-binding energy efficiency targets. It seems that the lessons of poor policy implementation of the EPBDs and the delays this means for the achievement of climate policy objectives have yet to be learned and applied in the EU's overarching policy on the promotion of energy efficiency.

Note

1   This is the average expected level across the five BAU scenarios presented by the European Commission in its 2050 energy roadmap (European Commission, 2011c).

References

Boasson, E. L. and Dupont, C. (2015). Buildings: Good Intentions Unfulfilled. In C. Dupont and S. Oberthür (eds), *Decarbonization in the European Union: Internal Policies and External Strategies* (pp. 137–158). Houndmills: Palgrave Macmillan.

Boasson, E. L. and Wettestad, J. (2013). *EU Climate Policy: Industry, Policy Innovation and External Environment*. Farnham: Ashgate.

BPIE. (2011). *Europe's Buildings Under the Microscope: a Country-by-Country Review of the Energy Performance of Buildings*. Buildings Performance Institute Europe.

CANEurope. (2011). *Energy Efficiency and Savings: Clearing the Fog. Everything You Always Wanted to Know but were Afraid to Ask about Europe's First Energy Source*. Brussels: Climate Action Network – Europe.

Council of the European Union. (1986). Resolution Concerning New Community Energy Policy Objectives for 1995 and Convergence of the Policies of the Member States. 15/16.IX.86.

Council of the European Union. (2001). 2347th Council Meeting: Energy/Industry. 8538/01 *(Presse 181)*.

Council of the European Union. (2002). Common Position (EC) No. 46/2002 Adopted by the Council on 7 June 2002 with a View to the Adoption of a Directive 2002/…/EC of the European Parliament and of the Council of … on the Energy Performance of Buildings. OJ C 197, 20, 6–15.

Council of the European Union. (2010). Position (EU) No. 10/2010 of the Council at First Reading with a View to the Adoption of a Directive of the European Parliament and of the Council on the Energy Performance of Buildings (Recast). OJ C 123, 32–58.

Dawson, B. and Spannagle, M. (2009). *The Complete Guide to Climate Change*. London: Routledge.

ECF. (2010). *Roadmap 2050. A Practical Guide to a Prosperous, Low Carbon Europe*. Brussels: European Climate Foundation.

EEA. (2010). *The European Environment: State and Outlook 2010 Synthesis*. Copenhagen: European Environment Agency.

EEA. (2012). *Consumption and the Environment – 2012 Update*. Copenhagen: European Environment Agency.

EEA. (2013). *Achieving Energy Efficiency Through Behaviour Change: What Does it Take? Technical Report No. 5/2013*. Copenhagen: European Environment Agency.

EEA. (2014). *Trends and Projections in Europe 2014. Tracking Progress Towards Europe's Climate and Energy Targets for 2020*. Copenhagen: European Environment Agency.

ENDS Europe. (2001a). MEPs Defy Ministers over Buildings Efficiency, *19 December 2001*.

ENDS Europe. (2001b). Ministers Neuter EU Building Efficiency Drive, *5 December 2001*.

ENDS Europe. (2002). MEPs Go Soft on Buildings Energy Efficiency, *13 September 2002*.

ENDS Europe. (2004). Insulation Firms Urge Wider Building Energy Law, *5 March 2004*.

ENDS Europe. (2005). Plea for Broader EU Buildings Energy Law, *16 March 2005*.

ENDS Europe. (2009a). All New Buildings to be 'Near-Zero-Energy' by 2020, *18 November 2009*.

ENDS Europe. (2009b). Council to Clash with MEPs over Efficient Buildings, *2 June 2009*.

ENDS Europe. (2009c). EU States Raise Doubts over Greener Building Plans, *16 February 2009*.

ENDS Europe. (2009d). MEPs Demand Financing Focus in Buildings Plan, *20 January 2009*.

ENDS Europe. (2009e). MEPs Set 2019 Deadline for Zero-Energy Buildings, *23 April 2009*.

ENDS Europe. (2009f). States' Fears for Green Building Law Revision Grow, *3 July 2009*.

ENDS Europe. (2010a). EU Risks Missing Efficiency Goal, Report Warns, *16 September 2010*.

ENDS Europe. (2010b). Financing for Greener Buildings 'Still an Issue', *29 February 2010*.

ENDS Europe. (2010c). MEPs Want Efficiency Focus in EU Energy Plan, *29 June 2010*.

EREC. (2010). *RE-thinking 2050: a 100% Renewable Energy Vision for the European Union*. Brussels: European Renewable Energy Council.

European Commission. (1979a). New Lines of Action by the European Community in the Field of Energy Saving. COM(79) 312.

European Commission. (1979b). Third Report of the Community's Programme for Energy Saving. COM(79) 313.

European Commission. (1984). Towards a European Policy for the Rational Use of Energy in the Building Sector. COM(84) 614.

European Commission. (1991). Proposal for a Council Directive on the Indication by Labelling and Standard Product Information of the Consumption of Energy and Other Resources of Household Appliances. COM(1991) 285.

European Commission. (1992). Proposal for a Council Directive to Limit Carbon Dioxide Emissions by Improving Energy Efficiency (SAVE programme). COM(92) 182.

European Commission. (1998). Energy Efficiency in the European Community – Towards a Strategy for the Rational Use of Energy. COM(1998) 246.

European Commission. (2000). Action Plan to Improve Energy Efficiency in the European Community. COM(2000) 247.

European Commission. (2001). Proposal for a Directive of the European Parliament and of the Council on the Energy Performance of Buildings. COM(2001) 226.

European Commission. (2002a). Amended Proposal for a Directive of the European Parliament and of the Council on the Energy Performance of Buildings. COM(2002) 192.

European Commission. (2002b). Communication from the Commission on Impact Assessment. COM(2002) 276.

European Commission. (2006). Action Plan for Energy Efficiency: Realising the Potential. COM(2006) 545.
European Commission. (2008a). Accompanying Document to the Proposal for a Recast of the Energy Performance of Buildings Directive (2002/91/EC): Impact Assessment. SEC(2008) 2864.
European Commission. (2008b). Proposal for a Directive of the European Parliament and of the Council on the Energy Performance of Buildings (Recast). COM(2008) 780.
European Commission. (2008c). Second Strategic Energy Review: an EU Energy Security and Solidarity Action Plan. COM(2008) 781.
European Commission. (2011a). Communication from the Commission. A Roadmap for Moving to a Competitive Low Carbon Economy in 2050. COM(2011) 112.
European Commission. (2011b). Executive Summary of the Impact Assessment Accompanying the Document: Energy Roadmap 2050. SEC(2011) 1566.
European Commission. (2011c). Impact Assessment Accompanying the Document: Energy Roadmap 2050. SEC(2011) 1565 Part Two.
European Commission. (2013). Commission Staff Working Document Accompanying the Document: Financial Support for Energy Efficiency in Buildings. SWD(2013) 143.
European Council. (2007). *Presidency Conclusions*, March 2007. Brussels: Council of the European Union.
European Council. (2014). *Conclusions*, Document EUCO 169/14, October 2009. Brussels: European Council.
European Parliament. (2001). Report on the Proposal for a Directive of the European Parliament and of the Council on the Energy Performance of Buildings. A5-0465/2001.
European Parliament. (2002a). Position of the European Parliament Adopted at First Reading on 6 February 2002 with a View to the Adoption of European Parliment and Council Directive .../.../EC on the Energy Performance of Buildings. 2001/0098(COD).
European Parliament. (2002b). Recommendation for Second Reading on the Council Common Position with a View to the Adoption of a Directive of the European Parliament and of the Council on the Energy Performance of Buildings. A5-0297/2002.
European Parliament. (2009a). European Parliament Legislative Resolution of 23 April 2009 on the Proposal for a Directive of the European Parliament and of the Council on the Energy Performance of Buildings (Recast). P6_TA(2009)0278.
European Parliament. (2009b). Report on the Proposal for a Directive of the European Parliament and of the Council on the Energy Performance of Buildings (Recast). A6-0254/2009.
Eurostat. (2015). Eurostat Online Database. Available at: ec.europa.eu/eurostat/data/database.
Haverland, M. and Romeijn, M. (2007). Do Member States Make European Policies Work? Analysing the EU Transposition Deficit. *Public Administration*, 85(3), 757–778.
Heaps, C., Erickson, P., Kartha, S. and Kemp-Benedict, E. (2009). *Europe's Share of the Climate Challenge: Domestic Actions and International Obligations to Protect the Planet*. Stockholm: Stockholm Environment Institute.
Henningsen, J. (2011). Energy Savings and Efficiency. In V. L. Birchfield and J. S. Duffield (eds), *Towards a Common European Union Energy Policy: Problems, Progress, and Prospects* (pp. 131–141). New York: Palgrave Macmillan.
IEA. (2011). *World Energy Outlook 2011*. Paris: OECD/International Energy Agency.

Lampinen, R. and Uusikylä, P. (1998). Implementation Deficit – Why Member States do not Comply with EU Directives. *Scandinavian Political Studies*, *21*(3), 231–251.

Liefferink, D., Arts, B., Kamstra, J. and Ooijevaar, J. (2009). Leaders and Laggards in Environmental Policy: a Quantitative Analysis of Domestic Policy Outputs. *Journal of European Public Policy*, *16*(5), 677–700.

Oberthür, S. and Pallemaerts, M. (2010). The EU's Internal and External Climate Policies: an Historical Overview. In S. Oberthür and M. Pallemaerts (eds), *The New Climate Policies of the European Union: Internal Legislation and Climate Diplomacy* (pp. 27–63). Brussels: VUB Press.

Oberthür, S. and Roche Kelly, C. (2008). EU Leadership in International Climate Policy: Achievements and Challenges. *International Spectator*, *43*(3), 35–50.

Richter, B. (2010). *Beyond Smoke and Mirrors: Climate Change and Energy in the 21st Century*. Cambridge: Cambridge University Press.

Roosa, S. A. (2010). *Sustainable Development Handbook*. Lilburn, GA: The Fairmont Press.

Wesselink, B., Harmsen, R. and Eichhammer, W. (2010). *Energy Savings 2020: How to Triple the Impact of Energy Saving Policies in Europe*. Brussels: Ecofys and Fraunhofer ISI.

5 EU policy on natural gas import infrastructure

In this chapter, I explore climate policy integration into the policy process and output of the EU's policies to support natural gas importing infrastructure. This policy area is embedded in the EU's external energy policy and aims to ensure secure supplies of natural gas to the EU, which has limited reserves of natural gas on its own territory. I focus particularly on policies that promote the construction of natural gas importing infrastructure (hereafter 'external gas infrastructure') connecting the EU to natural gas producing countries. Most notably, these policies include the EU's trans-European networks for energy (TEN-E) policies and the Regulation on a European Energy Programme for Recovery (EEPR). Before turning to a detailed discussion of these policies, however, I describe the role of natural gas in the EU up to 2050. I close the chapter with a short overview of potential future developments.

Natural gas in the EU

Natural gas is used in the EU mainly for heating, for some industrial processes, transport and, increasingly, for electricity generation (European Commission, 2010e; IEA, 2011). It is considered the 'cleanest' fossil fuel, with about half the GHG emissions of coal. It is relatively abundant globally. The already proven and in-place technology and infrastructure to extract and distribute gas make it an attractive option for the transition to decarbonisation. Its potential use as a 'transition fuel' (or bridging fuel) has resulted in much discussion on the benefits and disadvantages of moving towards more gas-based electricity generation instead of coal (IEA, 2011; Stephenson, Doukas and Shaw, 2012). Nevertheless, natural gas is still a fossil fuel and its continued use in the energy system cannot be reconciled with decarbonisation ambitions to 2050 unless it is coupled with CCS technology, which has yet (as of 2015) to be adopted to a sufficient scale to make a contribution to decarbonisation (Dupont and Oberthür, 2015b; Reichardt, Pfluger, Schleich and Marth, 2012).

Natural gas can be extracted and transported in different ways. Conventional natural gas is found in large deposits in gas fields and is extracted using drilling methods. It is generally transported by pipeline (unless it is converted to liquefied natural gas, LNG). LNG is natural gas that has been compressed and liquefied for transportation by ship. Shale gas, or 'unconventional' gas, is extracted from shale rock formations using a method called hydraulic fracturing (or fracking) that involves pumping a mix of water and chemicals underground to fracture the rock to release and capture the gas deposits within these rock formations. In the EU context, piped gas and LNG are the two main sources of natural gas for which importing infrastructure is required.

Natural gas consumption accounted for about 23 per cent of the EU's final energy consumption in 2013 (Eurostat, 2015). The EU consumed 438 billion cubic metres (bcm) of natural gas, but it produced only 147 bcm domestically (BP, 2014: 22–23). Hence, the EU is reliant on imports of natural gas to bridge the large gap between its production and consumption. Domestic production of natural gas has been steadily declining in the EU, from 225 bcm in 2003 (BP, 2014: 22). Consumption has been fluctuating over the same time period, but we see a slight downward trend in the EU also, especially since 2010. Highs of consumption were recorded for 2005 (497 bcm) and 2010 (502 bcm), but figures for 2013 show the lowest levels of consumption of natural gas in the EU over the previous ten years at 438 bcm (BP, 2014; Dupont, 2015).

Although natural gas consumption in the EU may be fluctuating or dropping, the shortfall between the levels of production and consumption nevertheless make securing imports of natural gas a necessity. When it comes to natural gas, therefore, the major policy concern seems to be securing supplies of natural gas. This focus was compounded by a series of crises in gas supplies in the 2000s. The EU was reliant on Russia for 39 per cent of its natural gas imports or 27 per cent of its gas consumption in 2013 (European Commission, 2014), with 15 per cent reaching the EU through transit pipelines in Ukraine (IEA, 2014). When Russia prevented flows of gas to Ukraine in 2006 and again in 2009, this had knock-on downstream effects on several EU member states. The crisis in Ukraine since 2013 has emphasised the need for the EU to seek alternative sources of energy beyond Russia, pushing energy security up the political agenda once more. A number of member states are almost completely reliant on Russia for their imports of natural gas, including Bulgaria, Estonia, Finland, Slovakia, Latvia and Lithuania (European Commission, 2014). These political and energy crises and realities have emphasised policies that promote infrastructure for accessing supplies of gas from alternative partners and for improving storage capacity and internal EU gas infrastructure connections. Beyond gas from Russia, EU member states also imported their natural gas from Algeria, Libya and Norway by pipeline and from Algeria, Egypt, Nigeria, Norway, Oman, Peru, Qatar, Trinidad and Tobago and Yemen by LNG in 2013 (BP, 2014).

EU natural gas infrastructure and policy

Natural gas imports rely on infrastructure – whether LNG terminals or pipelines. Although the EU has long-term objectives to reduce its GHG emissions by 80 to 95 per cent by 2050, infrastructure for importing natural gas is expanding. Additionally, such infrastructure projects generally have lifetimes of 50 years or more, so any pipelines or LNG terminals constructed since 2000 are likely to still be operational in 2050 (or will at best be poor investments and stranded assets) (Dupont, 2015; European Commission, 2010a).

According to the European Commission's Energy Market Observatory data, EU import pipeline capacity totalled about 441 bcm by 2012, with pipeline connections to Algeria, Russia, Norway and Libya. LNG capacity stood at about 181 bcm by 2012. LNG terminals exist in Belgium, France, Greece, Italy, Spain, Portugal and the United Kingdom (BP, 2014; European Commission, 2010b, 2010c, 2010d, 2011b, 2011c, 2012).

Several pipeline and LNG terminal projects are planned or under construction as of 2014. These include pipeline projects from Norway, Russia and the Caspian Sea. These pipelines would increase the infrastructure capacity from 441 bcm in 2012 to about 479 bcm per year by 2022, when all the planned projects are expected to be operational (Dupont, 2015; European Commission, 2014). Several existing LNG terminals plan to expand their capacity and a number of new terminals are either under construction or in planning phases (including in Spain, Italy, France, Poland and in the Baltic region). LNG terminal capacity is therefore expected to expand to about 259 bcm by around 2022 (Gas LNG Europe, 2014). Total gas infrastructure import capacity in the EU, therefore, is *increasing*. Adding the current capacity to the planned and under construction pipeline and LNG capacity that is due to become operational by about 2022, sees a total potential gas import infrastructure capacity in the EU of around 738 bcm in 2020 (see Table 5.1). Considering that levels of gas consumption in the EU stood at 438 bcm (some of which was domestically produced), one could wonder why such excessive import capacity is deemed necessary.

EU policy on importing energy infrastructure falls under mixed energy competence. Title XXI on energy of the Lisbon Treaty (TFEU, Art. 194) provides a clear mandate for the EU in the area of energy infrastructure development. Article 194 of the TFEU reads as follows:

1   In the context of the establishment and functioning of the internal market and with regard for the need to preserve and improve the environment, Union policy on energy shall aim, in a spirit of solidarity between member states, to:
    (a) ensure the functioning of the energy market;

Table 5.1 EU gas import capacity circa 2012 to 2022, taking account of projects under construction, but excluding proposed projects (measured in bcm)

	c. 2012	Increases by	c. 2022
Pipeline capacity	441	38	479
LNG capacity	181	78	259
Total	622	116	738

Source: Compiled from (Energy Observatory Market Data; European Commission, 2014; Gas LNG Europe, 2014; IEA, 2012)

130 *Natural gas import infrastructure*

      (b) ensure security of energy supply in the Union;
      (c) promote energy efficiency and energy saving and the development of new and renewable forms of energy;
      (d) promote the interconnection of energy networks.

Since the Lisbon Treaty entered into force in December 2009, therefore, the EU has a clear competence to develop policy on energy infrastructure. Yet this competence remains shared with the member states in that each state retains the right to choose their own energy mix. Article 192 states: 'such measures shall not affect a member state's right to determine … the general structure of its energy supply.' Policy on natural gas infrastructure decided before the entry into force of the Lisbon Treaty did not benefit from an energy chapter in the Treaties. Energy policy development in the EU was often made under environmental or internal market competences.

    The development of EU energy infrastructure policy can be linked to the objectives of ensuring sustainable energy development (linking RE generation to the grid and updating the grid to handle variable energy supply, for example), competitiveness (by ensuring choice of energy supplier and type of energy) and security of energy supplies (linking to new sources of energy supply, increasing internal interconnections, developing new routes for energy exchange) – the three objectives of EU energy policy overall (European Commission, 2008b: 2). In terms of sustainability objectives, it became clear that infrastructure developments would need to be developed hand-in-hand with RE generation (Sattich, 2015). This infrastructure development was required for three main reasons. First, much RE is generated in areas that are remote from the load centres where the energy is consumed (such as coastlines, mountains, offshore). Second, RE has long been considered hazardous for the old electricity grid infrastructure in place in the EU. As RE generation is dependant on hours of sunshine and wind power (as the two predominant RE sources today), the grid needed to cope with variable supplies at times during the day and year that do not necessarily correspond with peaks of consumption. Third, the potential for major sources of different types of RE is dispersed geographically throughout the EU. Much potential for wind power generation, for example, exists in the North Sea, and in countries such as Denmark, Scotland and Ireland, among others. Solar power potential is greatest in the Southern countries of the EU. Bringing this energy to the places where it is consumed within the EU requires large cross-border infrastructure (European Commission, 2010a; Sattich, 2015).

    While sustainable energy ambitions have promoted internal EU electricity infrastructure policy, it is more the energy security objectives of EU policy that have promoted external natural gas infrastructure. The European Commission's green paper in 2000 entitled a 'European strategy for the security of energy supply' highlighted the need for a 'stronger mechanism … to build up strategic stocks and to foresee new import routes' of natural gas supplies (European Commission, 2000: 4). Policy measures on stocks and emergency responses to disruptions in supply were implemented with Directive 2004/67/EC and the follow-up Regulation (EU) No. 2994/2010 on measures to safeguard security of natural gas supply. As they focus on short-term security measures rather than infrastructure development, I do not focus on these pieces of legislation here. Two policy measures in the 2000s directly provided support for external gas infrastructure, however, namely the trans-European networks for energy (TEN-E) guidelines (Decision 2003/1229 repealed by Decision 2006/1364), and the later policy on the European Energy Programme for Recovery (Regulation (EC) No. 663/2009, amended by Regulation (EU) No. 1233/2010).

Climate Policy Integration into EU Gas Import Infrastructure Policy

As already done in Chapters 3 and 4, I measure the level of CPI in the policy process in reference to the involvement of internal and external pro-climate stakeholders in the policy development and by seeking out evidence that policymakers recognise the functional interrelations between their policy domain and long-term climate policy objectives. Therefore, very high levels of CPI in the policy process should see involvement from traditional pro-climate stakeholders (such as DG Environment or DG Climate Action, the ENVI committee in the Parliament, the environment Council formation and external environmental and climate NGOs and industries). Furthermore, policymakers should demonstrate evidence (including articulation) that they recognise how energy infrastructure policies interact with climate policies, leading to climate-friendly policy outputs. In the particular case of natural gas infrastructure policy, the recognition of the functional interrelations and very high involvement of pro-climate stakeholders could conceivably lead to the abandonment of new policy in support of external gas infrastructure, as policymakers recognise that enough infrastructure is in place considering climate policy objectives should lead to decreases in consumption of all fossil fuels (including natural gas).

For the policy output, measuring very high levels of CPI means taking into account three points: (1) the expected natural gas consumption in a decarbonised EU in 2050; (2) the role to be played by CCS technology; and (3) the amount of gas infrastructure in 2050. By focusing on these three elements, it should be possible to establish what policy output for EU external gas infrastructure policy would be in harmony with the EU's long-term climate goals to 2050. From this benchmark, it is possible to measure how far policies already agreed in the EU bring us towards the 2050 climate goal.

First, with regard to consumption, decarbonisation scenarios to 2050 vary as to the amount of natural gas that will be required by 2050. Many scenarios include some amount of gas as a back-up fuel for intermittency of renewably generated electricity (ECF, 2010; European Commission, 2011d). Some scenarios claim that as little as 52 bcm of natural gas will be required in 2050, and only for industrial processes (Heaps, Erickson, Kartha and Kemp-Benedict, 2009), while electricity and heating will be completely decarbonised by 2050 (EREC, 2010). Eurogas promotes natural gas as a low-carbon energy source that could still provide as much as 462 bcm in 2050, with CCS (Eurogas, 2011). The

Commission's energy roadmap to 2050 outlines a range of estimates for gas consumption in its 2050 decarbonisation scenarios from 233 bcm to 320 bcm. In these scenarios, a minimum of approximately 202 bcm is used for gas-fired power generation, representing in each scenario the majority share of the use for natural gas in 2050 (European Commission, 2011d: 68–77). It is reasonable, however, to expect natural gas consumption to rise slightly in the short-term before beginning to decline as more coal plants are closed on the road to decarbonisation. Additionally, since EU domestic gas production is decreasing, most natural gas consumption in 2050 will come from imports (excluding the potential for shale gas production in the EU, which is a discussion that goes beyond the scope of this chapter).

Second, we must consider the role of CCS. None of the scenarios that foresee an amount of gas in the energy mix to 2050 can be compatible with long-term climate policy objectives without CCS, as full decarbonisation of the energy sector is required to achieve 80 to 95 per cent GHG emission reductions by 2050. The assumptions about the future of natural gas use in the energy mix to 2050 depend on the deployment of this technology for fossil-fuel power generation. Shell's 2050 scenarios outline that 90 per cent of all coal- and gas-fired power stations in the OECD would need to be equipped with CCS technologies in 2050 (Shell International BV, 2008: 32). The Commission's 2050 energy roadmap makes the assumption that CCS will play a role in between 19 and 32 per cent of total power generation by 2050, depending on other assumptions such as shares of RE generation (European Commission, 2011a: 8), yet it states earlier in the same document that it remains 'impossible to anticipate … whether and when carbon capture and storage (CCS) will become commercial' (p. 3). A 2012 appraisal of the state of the art of CCS technology outlines that it is not developing fast enough. Its role in 2050, under the current policy framework and current levels of development, would be 'marginal at best' (Reichardt et al., 2012: 1). Informal assessments with interviewees suggest that CCS is becoming less and less feasible as time goes on, with suggestions that companies are no longer interested in CCS (interviews 17 and 19). For the purposes of establishing the ideal level of CPI in the policy output, therefore, we cannot assume there will be a great roll-out of CCS technology, and it is the levels of natural gas consumption that must drop dramatically in order to achieve 2050 climate objectives.

Third, the amount of natural gas import infrastructure expected to be in place in 2050 should be considered when it comes to assessing the level of CPI in external gas infrastructure policy. As outlined above, the capacity for natural gas imports in 2022 is likely to amount to about 738 bcm. This adds 116 bcm of natural gas import capacity between 2012 and 2022 (see Table 5.1). As infrastructure is generally expected to have a lifetime of about 50 years (European Commission, 2010a), this 116 bcm of capacity will certainly still be in place in 2050, and even beyond. In addition, most EU gas import infrastructure is relatively young and much of the existing infrastructure is also likely still to be in place in 2050 (European Parliament, 2009). An estimate of available gas infrastructure in the EU to 2050 can thus be between 600 and 800 bcm, considering the infrastructure in place in 2012, and the infrastructure that should be ready by 2022. It should be noted that this estimate does not take account of proposed projects that could potentially be built over the next decade and remain operational beyond 2050. With investments in such infrastructure usually involving large sums (e.g. the Nord Stream pipeline connecting Russia and Germany is estimated to have cost nearly €8 billion to construct), there is a high economic pressure to continue operations for economic return, representing a high risk of 'lock-in' to this natural gas/carbon infrastructure.

Taking these three considerations (expectations of gas consumption under decarbonisation scenarios; the role of CCS; available import infrastructure in 2050), a very high level of CPI in EU external gas infrastructure policy output would involve decisions to limit expansion of infrastructure in line with efforts to reduce the consumption of natural gas (as a fossil fuel) to a minimum. Such a minimum level of consumption can be considered as including the consumption of sectors outside power generation, where substitution of substances may be difficult. Thus, taking account of expectations for natural gas consumption in 2050 in gross inland consumption, excluding gas for power generation, and considering gas use in, for example, industry, would lead us to a maximum figure of 52 bcm in natural gas consumption in 2050 – a rather radical scenario (European Commission, 2011d; Heaps et al., 2009). This figure excludes Eurogas' scenario of a major role for natural gas in power generation. Yet, current knowledge of the feasibility of a fully renewable energy generation system points to risks of intermittency until storage technology is better developed (Giordano, Gangale, Fulli and Sanchez Jimenez, 2011; Sattich, 2015; von Hirschhausen, 2010). To take a more conservative stance on the role of gas in 2050 would thus include a certain amount of gas-fired power generation. For the purposes of the analysis here, I therefore suggest up to a maximum of 150 bcm of natural gas may be required by 2050, based on a survey of scenarios to 2050, and provided the limited number of gas-fired power generation plants are equipped with proven CCS technology. Figure 5.1 compares such a CPI scenario with that of the BAU scenarios from 2005 and decarbonisation scenarios outlined by the European Commission's energy roadmap. This figure clearly shows the distance between very high levels of CPI and the Commission's decarbonisation scenarios.

However, the argument that natural gas may play a role in the short-term as an alternative to coal is valid. Figure 5.2 shows that a continuous level of natural gas consumption to about 2020 is feasible (as we move away from coal plant power generation, and potentially moving away from nuclear generation also, considering the post-Fukushima context). Such a reality will not, however, take away from the very high levels of CPI that need to be achieved before 2050, which remains at a natural gas consumption level of a maximum of 150 bcm. The effort in moving away from natural gas from 2015/2020 onwards becomes more pressing, with steeper consumption reduction ambitions required in later decades.

With natural gas consumption levels to drop under a decarbonisation scenario to a maximum level of 150 bcm (provided CCS technology can mitigate associated emissions), the growth in gas infrastructure seems absurd.

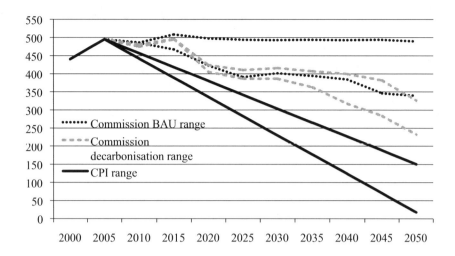

Figure 5.1 Range of very high CPI gas consumption trajectories in the EU to 2050, compared to the BAU and decarbonisation scenarios of the Commission's Energy Roadmap (measured in bcm)

Source: Compiled from (European Commission, 2011), own calculations; Mtoe converted to bcm using BP conversion rate, with 1 Mtoe = 1.11bcm

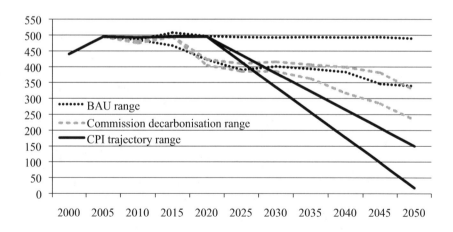

Figure 5.2 Very high levels of CPI to 2050, with continued natural gas consumption to 2020 before consumption begins to reduce in line with decarbonisation scenarios (measured in bcm)

Source: Compiled from (European Commission, 2011), own calculations; Mtoe converted to bcm using BP conversion rate, with 1 Mtoe = 1.11bcm

Infrastructure increases beyond 2020 are already providing for more capacity than required (see Table 5.1), and as gas consumption declines in the decades to 2050, such infrastructure represents an investment risk. Worse, the infrastructure capacity could delay the transition to decarbonisation as sunk investments call for returns. Taking all this into account, we can argue that EU external gas infrastructure policy should rather discourage investment in new importing infrastructure if it is to be in line with 2050 climate objectives. This does not prevent the EU from improving its internal infrastructure connections, but promoting *external* infrastructure to get more gas from more partners externally does not seem to be in harmony with climate policy objectives.

EU natural gas import infrastructure policy in the 2000s

There are two main policies that were agreed in the EU in the 2000s that aim directly to promote natural gas importing infrastructure – the TEN-E guidelines and the EEPR. I analyse these two policies in particular for evidence of climate policy integration in the next section. Here, I discuss the development of TEN-E Decisions and the EEPR Regulation in the context of overarching policy for securing supplies of natural gas in the EU.

EU policy on the promotion of external natural gas infrastructure is most closely linked to objectives and discussions on improving the security of supply (Bahgat, 2006; Baran, 2007; Brown and Huntington, 2008; Dupont, 2015; Skea, Chaudry and Wang, 2012). Energy security has been one of the major motivating factors for the development of EU energy policy since the very beginnings of the European integration project. In its 1995 green paper on energy policy, the Commission outlined the 'trinity' of EU energy policy goals that have remained central to EU energy policy since: security of supply, competitiveness, and sustainability (European Commission, 1995: 5). While policies to promote external gas infrastructure connecting third countries to the EU thus ought to consider the implications of each of these three policy objectives, the measures are situated most directly in the discourse on the security of supply.

With EU natural gas production steadily declining, while consumption declines more slowly, there has been an overall increase in the dependence of the EU on imports of natural gas, with Russia being the main supplier to the EU. The EU was 65 per cent dependent on imports of natural gas in 2015, compared to 62 per cent in 2010 and 57 per cent in 2005 (Eurostat, 2015). The increasing dependence on external suppliers was not lost on policymakers in the EU, with the adoption of Council Directive 2004/67/EC concerning 'measures to safeguard security of natural gas supply' in 2004. This Directive promoted measures for heightened security such as: improving pipeline capacity internally in the EU; improving domestic production of gas; diversifying sources of gas supply; and investing in pipelines and LNG terminals for importing gas. Fears of weaknesses in energy supply security seemed to be realised in 2006 and 2009 when supplies of gas from Russia were cut for periods of time due to disputes with Ukraine (a country that transits Russian gas onwards to the EU).

Generally, the measures outlined in the Directive focus on measures within the capacity and competence of the EU at that time, namely by encouraging member state solidarity in creating (physical) links in the internal energy market. A strong emphasis on subsidiarity and proportionality means that a three-step response approach was agreed under the Directive. First, industry responds to supply interruptions with their own measures. Next, member states step in if industry cannot respond adequately in cases of supply disruptions. Only if member states cannot sufficiently respond to the situation will the EU step in as a third-level measure.

In 2008, the Commission reported on the implementation of the 2004 Directive (European Commission, 2008a). The 2006 interruption of gas supplies was the most severe gas supply crisis to have taken place since the adoption of the Directive. This crisis was managed effectively by measures in place at the national level, and EU-level emergency responses were not required. The crisis did mean that the EU Gas Coordination Group established by the 2004 Directive (Art. 7) met for the first time (European Commission, 2008a: 6). Given the brevity of the crisis (36 hours) and the adequacy of national measures to respond to the situation, by the time the Gas Coordination Group met, full supplies of gas had already been restored (p. 7). The difficulty in ensuring a sufficiently quick emergency response to sudden interruptions in gas supplies at the EU level indicated some inadequacy in the measures laid out in the 2004 Directive. This was further compounded by the more severe 2009 gas crisis.

Another important moment in the development of EU energy security policy came with the second strategic energy review. The Commission released the second strategic energy review, subtitled 'an EU security and solidarity action plan', in November 2008 (European Commission, 2008b). This communication is often considered a watershed moment in the development of EU security of supply policy (interviews 22 and 24). It followed the 2005 informal meeting of heads of state and government at Hampton Court in the UK that emphasised increased EU coordination in energy policy. It called especially for promoting further energy infrastructure to meet the EU's energy needs (European Commission, 2008b: 4), and gave special priority to supporting LNG projects and the Southern gas corridor initiative to bring supplies of gas from the Caspian Sea region (and potentially also from the Middle East) to Europe. One interviewee outlined that this particular communication and its acceptance by the Parliament and Council allowed for more ambitious infrastructure promotion to follow (including under the EEPR, see below, which provided an unprecedented amount of financial support to energy infrastructure projects in the EU; interview 24).

In January 2009, a dispute between Russia and Ukraine caused several weeks of disruptions in gas supplies to the EU. This crisis was labelled 'unprecedented' by the EU (European Commission, 2009a: 2; European Council, 2009: 9). It demonstrated major weaknesses and vulnerabilities in the emergency responses of the EU in the face of such severe supply disruptions. As a result, the EU agreed to a revision of the 2004 Directive in the shape of Regulation (EU) 994/2010 concerning measures to safeguard security of gas supply and repealing Directive 2004/67/EC. While the context of gas supply security in the EU revolves around three main issues – increasing import dependence; increasing dependence on Russia and transit routes; and vulnerability to gas supply interruptions – the 2004 Directive and the 2010 Regulation respond only in a short-term way to the third challenge. With the long-term perspective to 2050 decarbonisation, I focus in this chapter on the level of CPI in the EU's long-term policies to support external gas infrastructure projects.

The trans-European networks for energy (TEN-E) policy is one of the main instruments in the EU for supporting gas infrastructure. The TEN-E guidelines were first put in place in 1996, under Decision 1254/96/EC of the European Parliament and the Council. This Decision laid down guidelines for assigning the label of a 'project of common interest' to trans-European gas (and electricity) infrastructure projects, with the aim of: introducing natural gas to new regions; connecting isolated gas networks to European networks; and increasing the transmission, reception and storage capacity while also diversifying the sources and routes of natural gas imports (Art. 2.2). TEN-E policy benefits from a separate legal competence in the Treaties (Articles 170 to 172 of the Treaty on the Functioning of the EU).

In 2001, the Commission proposed an amendment to the 1996 TEN-E Decision (European Commission, 2001). The revision aimed to 'underline political support' for a special category of priority TEN-E infrastructure projects of common European interest (pp. 23, 40). The revision outlined 'criteria upon which support will be determined; notably the creation of the internal market, security of supply, integration of renewable energy, enlargement and the integration of ultra-peripheral regions' (p. 23). The Commission suggested gas connections directly between Russia and Germany (later realised as the Nord Stream pipeline, construction completed in 2012), connections between the Caspian Sea region and the EU (the 'Southern gas corridor'), and a gas pipeline connection between Algeria and Spain (construction of the Medgaz pipeline between Algeria and Spain was completed in 2008), as the three 'priority axes' of European interest to benefit from specific financial and political support (p. 25).

It may be interesting to pause here and provide some background information on two of these external infrastructure projects – namely Nord Stream and the Southern gas corridor. The North European Gas Pipeline (now known as Nord Stream) is a set of twin submarine pipelines in the Baltic Sea, connecting Vyborg in Russia to Greifswald in Germany. The project is considered to increase EU gas security by creating a direct link between the EU and Russia and bypassing transit countries. The first of the twin pipelines commenced operations in November 2011, while the construction of the second pipeline was completed in April 2012. Together, these pipelines have the capacity to carry 55 bcm of gas from Russia directly to Germany every year. They have an expected lifetime of about 50 years (until beyond 2060). From its early days (discussions on the pipelines began as far back as 1993), the Nord Stream project received unofficial political support from many EU member states (Larsson, 2007). With the 2003 revision of the TEN-E guidelines, the project received official EU-level political backing as a 'priority project' with opportunities for it to access financial support. The Nord Stream consortium did not, in the end, make use of the potential funds available under TEN-E, and is completely privately funded. However, the political support attached to the 'priority project' label certainly aided its realisation in terms of negotiations for financial backers and getting the go-ahead from Baltic states (interview 13).

The Southern gas corridor is an initiative that is designed to increase the EU's security of gas supply by diversifying suppliers. Instead of importing more gas from Russia, the EU-backed project sources gas from the Caspian Sea region, with the potential to expand later to source supplies from the Middle East. These regions have substantial reserves of natural gas. The main supplier countries proposed for the Southern gas corridor include Azerbaijan, with a proven gas reserve of about 1 trillion cubic metres (tcm) at the end of 2013, Turkmenistan with 17.5 tcm, Iraq with 3.6 tcm, and potentially Egypt (1.8 tcm) and Iran (33.8 tcm) in the future, political relations pending (BP, 2014: 20). This gas corridor would thus allow the EU access to some of the largest reserves of natural gas in the world (for comparison, Russia's gas reserves at the end of 2013 are listed as 31.3 tcm). The pipeline consortium that is expected to build a connection to the Caspian Sea region is the Trans-Adriatic Pipeline (TAP), which has received financial support under TEN-E, although this pipeline did not hold the priority project label.

The Commission's proposal for a revision of the TEN-E guidelines moved through two readings in the Parliament and the Council before being adopted as Decision 1229/2003 on 26 June 2003. In neither the Parliament nor the Council did internal environmental or climate actors play a visible role in the decision-making process. In the Parliament, the ITRE committee drafted the Parliament's first and second reading opinions, with the ENVI committee not considered for an opinion. In the Council, the final decision was taken in the general affairs meeting in June 2003, although the proposal had previously been discussed in the industry Council meeting in June 2002, the transport, telecommunications and energy Council meeting in November 2002 and also in the education, youth, culture and sport Council meeting in February 2003. There was no discussion in the environment Council meetings.

While the policy negotiations did lead to some changes between the proposed text and the final act, there was no change to the suggested external gas infrastructure projects (outlined in the Annexes to the Decision). Both the proposal and the final act highlighted gas pipelines between Algeria and Spain, between Russia and the EU, and from the Caspian Sea and Middle East to the EU, as well as LNG terminals in Belgium, France, Spain, Portugal and Italy as 'axes for priority projects' (Annex I). This meant that projects listed under Annex III as 'projects of common interest' that fall within the 'axes for priority projects' listed in Annex I 'shall have priority for the grant of Community financial aid' (Art. 7).

Another revision of TEN-E guidelines amended the 2003 Decision with Decision No. 1364/2006/EC. This revision was deemed necessary mainly to incorporate the newly acceded member states into the scope of the guidelines (Decision 1364/2006/EC, Recital 1). The 2006 Decision specified that the label 'project of European interest' could be applied to projects that are of a cross-border nature or have 'significant impact on cross-border transmission capacity' (Art. 8.1). These projects would receive priority in the selection process for funding under TEN-E, but they would also receive 'particular attention' under other Community funds (Art. 8.1 and 8.2). As listed in Annex I of the 2006 Decision, several external gas infrastructure projects were assigned the label 'project of European interest', including Nord Stream, pipelines connecting Algeria with Spain and Italy, and two proposed pipelines in the Southern gas corridor: the Turkey-Greece-Italy interconnector (ITGI) and the Turkey-Austria pipeline (Nabucco), and a pipeline connecting Libya to Italy (Annex I).

The process leading to the adoption of the 2006 revision of the TEN-E was initiated in December 2003, just six months after the adoption of the previous TEN-E Decision (European Commission, 2003b). In the Parliament, the ITRE committee was responsible for providing an opinion on the Commission's proposal, with MEP Anne Laperrouze (ALDE) as rapporteur. The committee presented its first reading report to the Parliament's plenary in May 2005. The ENVI committee in this case gave an opinion (MEP Claude Turmes of the Greens political group was the rapporteur for opinion). Interestingly, the ENVI opinion includes much criticism of the proposed long list of projects under the guidelines, due to a lack of stringent and clear long-term criteria. The ENVI committee proposed deleting the article referring to 'projects of European interest' based on the fact that the choice of such projects implies bypassing consultation and authorisation through local authorities (European Parliament, 2005: 69). This suggestion was not taken up in the ITRE committee rapporteur's final report. Yet when it comes to supporting gas infrastructure, the ENVI opinion is positive on the role of gas in combating climate change:

> A well functioning gas market needs regulation on access to gas storage. Europe needs also to regularly improve its gas efficiency with long-term instruments such as quantitative commitment by distributors to reduce consumption. Only such measures will allow the EU to fulfil its commitment on climate change.
>
> (ibid.: 60)

As this quote implies, the ENVI committee saw a continued role for gas in the EU energy mix, provided efficiency gains were realised, the internal market was upgraded and long-term measures to reduce gas consumption were adopted. Yet, there is no clear objection from ENVI to the support of the external gas infrastructure projects. In fact, gas is seen here as a component in the fight against climate change. This is understandable as a short-term solution to moving away from the more GHG-intensive use of coal. Yet the risk of 'carbon lock-in' from promoting gas in the energy system was not considered in the committee procedures on the TEN-E guidelines, and not even in the ENVI committee.

140   *Natural gas import infrastructure*

In the Council, the proposal was discussed three times in the transport, telecommunications and energy Council formation in 2004 and 2005, before being agreed upon in the competitiveness Council meeting (on internal market, industry, research and space) in 2006. The Council adopted its common position on 8 December 2005, and substantially altered the Decision. It deleted the articles relating to the promotion of the 'project of European interest' label and the nomination by the Commission of a European coordinator for such projects (Council of the European Union, 2005a). In the explanation of its position, the Council expressed concern that the use of a separate category of projects would result in a 'negative effect' on the realisation of other viable projects (Council of the European Union, 2005b). With regard to the appointment of a European coordinator, the Council argued that 'far less bureaucratic provisions could be retained for the same purpose' (Council of the European Union, 2005b: 4). Despite informal trialogue negotiations during the first reading, these issues could not be resolved. The Commission stated, in its response, that it could not accept the text as proposed by the Council (European Commission, 2005), and the file moved to second reading in the Parliament in January 2006. The Commission especially called on the Council to take note of the conclusions of the informal Hampton Court meeting of heads of state or government in October 2005, and quotes former UK Prime Minister Tony Blair, that: 'in respect to energy, there was an agreement to take forward work in the energy sector, including how we try to establish a common European grid. … it is important too that energy policy is something that we work on together as a European Union' (European Commission, 2005: 5).

The second reading report adopted by the Parliament was the result of compromise negotiations between the Council and the Parliament that reinserted the 'project of European interest' category and the 'European coordinator'. This time, the Commission could accept all (17) amendments proposed by the Parliament (European Commission, 2006: 3). As part of the compromise, and on member state insistence, the Commission is not be able to withdraw the declaration of European interest from a project. The second reading decision by the Parliament came on 4 April 2006 and the Parliament and the Council signed the final Decision on 6 September 2006.

As part of the EU's response to the 2008 economic crisis, the EEPR was agreed in 2009. This first Regulation (No. 663/2009) outlined the conditions and eligibility requirements for infrastructure projects related to gas, electricity, CCS and offshore wind to access funding directly from the EU. Such infrastructure projects were seen as needing specific financial help during a financial and economic crisis that saw little private investment. By providing the financial boost to help the projects go ahead, the aim was also to provide jobs. The EEPR applies concrete policy support for long-term security of natural gas supply in the form of promises of funding to certain external gas infrastructure projects. The Regulation thus ensures infrastructure is in place to meet the goals of diversifying gas importing routes and sources (Art. 4). Eligible projects were listed in the Regulation's Annex. In terms of external gas infrastructure, the Regulation earmarks funding for four external gas infrastructure projects (out of a total of 18 gas infrastructure projects amounting to €1.44 billion):

- €200 million for the proposed Nabucco pipeline project in the Southern gas corridor;
- €100 million for the proposed Italy-Turkey-Greece Interconnector (ITGI) – Poseidon pipeline project (also in the Southern gas corridor);
- €120 million for the Galsi pipeline connecting Algeria and Italy;
- And €80 million for an LNG terminal on the Polish coast (Regulation (EC) No. 663/2009, Annex).

The funding can be awarded to project promoters as grant funding, and can cover up to 50 per cent of the eligible investment costs for gas projects (as outlined in Art. 9). Together, these four projects would have the capacity to supply the EU with an additional 57 bcm of natural gas supplies per year.

The 2009 Regulation was proposed and adopted under Article 156 (on trans-European networks, now Articles 170–172 TFEU) and Article 175.1 of the environment chapter of the Treaty establishing the European Community (TEC). In the policy process leading to the adoption of the EEPR Regulation, it is particularly interesting to note that the Commission did not open a public consultation, nor did it prepare an impact assessment. This was considered justified due to the urgent need to respond to the economic crisis (and the fact that funds needed to be committed within the year). It was hence argued that there was insufficient time to complete a consultation process and develop an impact assessment (European Commission, 2009b: 2–3). In the Parliament, MEP Eugenijus Maldeikis (Union for Europe of the Nations political group) served as rapporteur in the industry, research and energy committee (ITRE). The environment committee (ENVI) decided not to give an opinion. In the Council, the proposal was discussed once in the energy Council, twice in the general affairs Council and twice (including for the final Council agreement) in the economic and financial affairs Council meeting. From 28 January 2009, when the Commission presented its proposal, less than six months passed before the final act was signed on 13 July 2009.

In its April 2010 review of the 2009 Regulation, the Commission expressed a general satisfaction with the implementation of the EEPR (European Commission, 2010f). The implementation procedure began with a call for proposals in the three sub-programmes of the EEPR (gas and electricity infrastructure, offshore wind and CCS) in May 2009, with 33 proposals for gas infrastructure projects submitted. After evaluation by the Commission and relevant committees (principally, the TEN-E financial assistance committee), the Commission decided to grant funding to 31 of the gas projects for a total of EUR 1.391 billion (European Commission, 2010f: 7). The Commission described with satisfaction that the EEPR, at this early stage in its implementation, seemed to be acting as a stimulus for attracting investors to the supported projects (p. 8). In terms of the external gas infrastructure projects originally listed in the Annex to the Regulation, the two pipelines singled out for financial assistance in the Southern gas corridor eventually did not go ahead, as the TAP became the Azeri's pipeline of choice for exporting their gas to the EU (Dupont, 2015). A follow-up Regulation to the 2009 EEPR Regulation was adopted in 2010 to assign any remaining funds to the development of energy efficiency and renewable energy initiatives (Regulation No. 1233/2010), but no further funding for gas infrastructure projects was assigned in this second Regulation.

CPI into the policy process of the TEN-E Decisions (2003 and 2006)

Internal pro-climate stakeholders

Throughout the negotiations on the TEN-E guideline revisions, leading to both the 2003 and to the 2006 Decisions, there is limited evidence of internal environmental or climate voices playing a role. Internally to the decision-making processes in the EU, DG Energy was in the lead in the Commission, with the ITRE committee drafting the first reading opinion report in the Parliament and various formations of the Council negotiating the texts, but notably never the environment Council. In the lead up to the 2006 TEN-E Decision, the ENVI committee in the Parliament did provide an opinion on the proposed text for the consideration of the ITRE committee. Their concerns were with the label 'project of European interest', and less with long-term concerns for carbon lock-in into natural gas infrastructure. Interestingly, the only real consideration of natural gas infrastructure in the ENVI committee's opinion to the ITRE committee was to promote efficient use of natural gas in the short term as part of the solution to combating climate change. While such a position may be understandable as gas represents an alternative to coal, the support of infrastructure represents a long-term strategy, with the infrastructure expected to operate for at least 50 years. Here, it is clear that even pro-climate actors in the EU either did not promote or could not promote long-term climate-friendly solutions that would see no support for further external gas infrastructure. Electricity interconnections received more attention from internal climate stakeholders than gas interconnections. As long as renewable energy was mentioned and highlighted in the TEN-E guidelines (which was the case), it seems the internal environmental voices were satisfied. In the case of the TEN-E revision process, pro-climate voices were not very active or visible in the policy process, and neither did they represent the long-term pro-climate position that would promote the abolition of political and financial support for fossil fuel infrastructure. As such, internal pro-climate stakeholder involvement is *very low* (with little to no involvement and visibility), as even when a traditional pro-climate voice was involved it did not promote CPI.

External pro-climate stakeholders

External pro-climate stakeholders were not visible nor were they involved in the policy process leading to the revisions of the TEN-E guidelines in 2003 and 2006. One interviewee reported that the main stakeholders consulted, both formally and informally, in both these policy processes were energy companies and gas and electricity network operators (interview 24). The consultations carried out by the Commission in the lead up to the TEN-E proposal published in 2001 included a questionnaire that was deemed to have reached all 'relevant parties' (European Commission, 2001: 4). These included member states, national energy regulatory authorities, network operators, gas and electricity industries, consumers, traders and other infrastructure users, and financial institutions. Renewable energy companies may be subsumed under 'electricity industries' and 'other infrastructure users' in this list, but there is no indication of input from environmental NGOs. For the 2006 TEN-E Decision, the Commission opened an online consultation between 25 July 2003 and 15 September 2003, which received 17 replies, with the majority of replies (13) referring to the electricity sector (European Commission, 2003a: 28), with no environmental NGOs or voices responding to the natural gas sector elements of the proposed Decision. In this case, the environmental interests did not question the broader environmental or climate impacts of supporting infrastructure projects. They were not focused on addressing support for external gas infrastructure. Concerns that promoting gas infrastructure would lock the EU into a carbon-based energy system were not raised.

The existence of TEN-E policy since the 1990s also meant that relations between policymakers and, especially, energy companies were well developed (interview 24; Vasileiadou and Tuinstra, 2013). External pro-climate stakeholders were not specifically targeted for consultation in TEN-E policy. Procedures for open consultation were in place for the revisions, so it is not a case that such stakeholders are deliberately excluded. It is more likely that they simply are not considered to be among the most relevant stakeholders in such policy processes. It seems that pro-climate stakeholders themselves did not seek to raise the long-term climate concerns in gas infrastructure policy either. They did not perceive this policy sector as a priority for their involvement, as they considered the greatest urgency to be a reduction in coal consumption (interviews 7 and 8). This is a case of stakeholders choosing their battles (Hauser, 2011), and policymakers working with their traditional partners. The lack of external pro-climate stakeholders in the policy process means that CPI is not promoted through them – their involvement is *very low* CPI.

Recognition of functional interrelations

The *very low* involvement of both internal and external pro-climate stakeholders in the TEN-E policy processes seems to be linked to the lack of any recognition or articulation in the policy process of the functional interrelations between long-term climate policy objectives and external gas infrastructure policy. The interrelations between external gas infrastructure and combating climate change play out on the long-term horizon. While it is reasonable to expect gas to play a role in the short-term as an alternative to coal, long-term climate policy requires a total decarbonisation of the EU's energy sector by 2050. This requires GHG emissions to peak sooner rather than later, meaning that adding new fossil fuel capacity is counter to decarbonisation aims. When gas consumption under a decarbonisation scenario needs to fall dramatically to 2050, it is clear that no further external gas infrastructure is required if decarbonisation objectives are to be achieved. The recognition of these interrelations in the policy discourse should lead policymakers to abandon political and financial support of external gas infrastructure. As described above, this was certainly not the case for the TEN-E revisions. Not only were these long-term climate policy objectives unrecognised in the policy process by policymakers, but also even the traditional pro-climate voices did not highlight these interrelations. Where climate policy objectives were mentioned, they were seen to benefit from further gas infrastructure as a short-term climate solution. With these infrastructure projects taking many years to be realised, and having lifetimes of 50 years or more, it is clear that the long-term functional interrelations were not recognised. Thus, this demonstrates *very low* recognition of functional interrelations.

In summary, the level of CPI in the policy process leading to the revisions of the TEN-E guidelines is *very low*. This is an aggregate of the *very low* levels of CPI evident through the role of internal and external pro-climate stakeholders in the policy process and the lack of recognition of functional interrelations between external gas infrastructure policy and long-term climate policy objectives.

CPI in the policy output of the TEN-E Decisions

As discussed above, abandoning policies that support new natural gas importing infrastructure would represent the highest level of CPI in policy output that can be expected. It is difficult to conceive of a policy output that provides political and financial support to several external gas infrastructure projects as demonstrating any levels of CPI. The rules governing the financial aid under TEN-E are laid down in Council Regulation (EC) No. 2236/95 on the 'granting of Community financial aid in the field of trans-European networks'. As stipulated in Article 4, aid for TEN-E projects can include co-financing of up to 50 per cent of the cost of preparatory, feasibility and evaluation studies, subsidising the interest on loans from the European Investment Bank or other financial entities, contributing towards fees to guarantee loans, and providing direct grants in certain cases. Generally, the budget of the TEN-E programme has been relatively modest, with an annual envelope of about €20–23 million (European Commission, 2003a: 9; interview 24). In 2003, this allowed funding for between 10 and 20 projects, with the funds being used mainly for the initial study phases of the projects (European Commission, 2003a: 11). The political importance of the labels for projects assigned under the TEN-E guidelines, however, should not be underestimated. Even in cases where no funding has been applied to a project (such as with Nord Stream), the label 'project of European interest' has helped in assigning it credibility and status (interview 13). Indeed, the Commission outlines that the indirect effects of the recognition of a project as one of 'European interest' may by more important for the project's development and advancement than any financial input (European Commission, 2003a: 11).

Political and financial support for energy infrastructure projects in general need not be in conflict with climate policy objectives, if those projects are linked to ensuring climate policy objectives are met (such as upgrading the electricity grid to handle increased renewable generation). The very existence of external gas infrastructure projects in the TEN-E guidelines, however, suggests that this part of the policy is in conflict with climate policy objectives. As such, CPI in the policy output of TEN-E guidelines on external gas infrastructure can only be measured as *very low* (see Table 2.2). The projects supported in the 2003 and 2006 TEN-E guidelines have had mixed levels of success. In terms of gas pipelines, since the agreement of the 2003 guidelines, construction has been completed on the 8 bcm Medgaz pipeline (2008) and the 55 bcm Nord Stream pipeline (2012). The supported Southern gas corridor pipelines – Nabucco and ITGI – were not chosen as the supply route for gas from Azerbaijan to the EU. But the infrastructure projects constructed (or planned) since 2000 will remain in place beyond 2050, when the EU should have already achieved its long-term decarbonisation goals. In this case, the TEN-E guidelines actually help widen the gap between the status quo and very high levels of CPI.

CPI into the policy process of the EU's EEPR Regulation (No. 663/2009)

Internal pro-climate stakeholders

With the involvement of internal pro-climate stakeholders (such as DG Environment, the ENVI committee in the Parliament, and the environmental Council) in the internal decision-making processes, there are more opportunities to advance CPI. In the case of the 2009 EEPR Regulation, however, these pro-climate stakeholders were missing from the policy process. In the Commission, DG Energy was in the lead on the policy proposal, and while they proposed support for offshore wind projects as well as the external gas infrastructure projects discussed above, at no point did any internal pro-climate stakeholder question the inclusion of gas infrastructure projects. There was no impact assessment procedure for the EEPR Regulation, which may also have limited cross-sectoral discussions on the Regulation within the Commission. In the Parliament, the ENVI committee decided not to give an opinion on the policy proposal, and in the Council, the proposal was not discussed in the environment Council formation. Furthermore, the speed of the decision-making process (less than six months) meant that little change was made to the Commission's original proposal. As 2009 was an election year in the Parliament, and as the proposal was sped through, the EEPR was not even discussed or negotiated in detail within the ITRE committee, let alone among different committees (interviews 25 and 26).

Under such circumstances, the pro-climate stakeholders in the EU decision-making processes were not involved and were not visible in the policy process leading to the 2009 EEPR Regulation, meaning their involvement is measured as *very low* (see Table 2.1).

External pro-climate stakeholders

The 2009 EEPR Regulation was proposed and negotiated under unusual circumstances for the ordinary legislative procedure. It was proposed by the Commission as a quick response to the economic and financial crises. The Commission thus did not carry out the usual consultation procedures. There was no open consultation process, and neither was there an impact assessment accompanying the proposal (European Commission, 2009b). Unlike in the case of the TEN-E revisions, the usual formal opportunities to engage external stakeholders were missing. The access of stakeholders was thus limited to later stages in the negotiations after the Commission had made its proposal, and to informal lobbying activities with the Parliament and Council.

Even so, pro-climate stakeholders were not visible and neither were they involved in the policy process leading to the 2009 EEPR Regulation. As with the TEN-E Decisions, this could be due to the lack of recognition of the functional interrelations between long-term climate policy objectives and external gas infrastructure; or it may be related to the lack of pro-climate stakeholder resources to get involved in every policy process. Beyond the limited access, especially in the early stages of the policy process, the speed of the negotiations in this case also left many pro-climate stakeholders outside the process (interviews 8 and 24). With Parliament insisting on including a Recital in the Regulation that unspent funds be used for energy efficiency and renewable energy projects (creating an energy efficiency fund and resulting in the 2010 EEPR Regulation, see above), external pro-climate stakeholders acted mostly to ensure that this obligation was fulfilled later. In the process leading to the 2009 EEPR they were neither involved nor visible in the process to the extent that they could raise levels of CPI in objection to the external gas infrastructure support provided by the EEPR. Their involvement is thus measured as *very low*.

Recognition of functional interrelations

As with the TEN-E revisions, it is plausible that the low involvement of internal and external pro-climate stakeholders is related to a lack of recognition (by all parties) of the functional interrelations between promoting external gas infrastructure and achieving long-term climate policy objectives. Where climate change was mentioned in the policy process, it was rather in reference to the other projects supported under the EEPR (namely, CCS and offshore wind projects), as they are seen as responding directly to the climate challenge. However, even within this one policy measure, there was no explicit questioning of the conflict between supporting external gas infrastructure projects while at the same time trying to achieve long-term climate policy objectives and decarbonisation. Even within a single policy measure, conflicts are present. Recognition of the functional interrelations could rather have led to a questioning of the need for further support for such infrastructure projects by policymakers or at least a reduction in the amounts of funding assigned to them. At best, external gas infrastructure projects could have been excluded from the EEPR. As the policy process unfolded, the Galsi pipeline was added to the original proposal, and external gas infrastructure projects were earmarked for €500 million, compared to €430 million in the Commission's proposal (European Commission, 2009b: 25). In the case of the external gas infrastructure support provided by the EEPR Regulation, there is very little evidence of the recognition of functional interrelations: a *very low* score.

CPI in the policy output of the 2009 EEPR Regulation

In many respects the EEPR Regulation builds on the policy measures of the TEN-E framework. The external gas infrastructure projects identified under EEPR had already received preliminary funding under TEN-E for their initial scoping investigations and/or were considered as 'projects of European interest' under TEN-E. While TEN-E provides support for projects in their early stages of development, EEPR targets more mature projects that are likely to need funding support to develop further under the context of the economic downturn. Additionally, the funding earmarked in TEN-E is limited. TEN-E makes available an annual financial envelope of about €20 million, whereas the EEPR earmarked €500 million for the four external gas infrastructure projects listed above, and a total of €1.44 billion for all gas infrastructure projects.

Very high levels of CPI would indicate that no support (financial or political) should be given to external gas infrastructure projects that are not needed under decarbonisation. At the time of the agreement on the EEPR, the Nord Stream pipeline was under construction and no longer required the political support provided by the TEN-E Decisions. Nord Stream creates a further 55 bcm of external gas infrastructure capacity (the twin pipelines were completed in 2012). The four projects supported by EEPR would add up to 57 bcm of further capacity of external gas infrastructure. These projects seem to have been inherited from the TEN-E Decisions, but the financial support made available under EEPR was substantially larger than anything available under TEN-E. The EEPR not only continues the misplaced political and financial support of external gas infrastructure projects that are not required under long-term climate policy objectives to decarbonise by 2050, but also actually increases the financial support. Thus, the EEPR widens the gap between very high levels of CPI in external gas infrastructure policy and the policy reality. CPI in the policy output of the EEPR policy on external gas infrastructure is thus *very low*. In fact, in both the TEN-E and EEPR cases, it could even be argued that we see 'negative' levels of CPI in the policy output.

Summary and outlook for 2030 and beyond

In this chapter, I examined CPI in the policy output and process of policies in support of external gas infrastructure between 2000 and 2010. These policies included explicit financial and political backing provided, at the EU level, through the 2003 and 2006 revisions of the TEN-E guidelines and the 2009 EEPR Regulation. In the discussion on the external gas infrastructure policy of the TEN-E guidelines and the EEPR Regulation, it was clear that very low levels of CPI were evident in these cases.

Since 2010, two policy developments in the EU have focused on further developing energy infrastructure – namely the 2013 Connecting Europe Facility (CEF) and plans for an 'Energy Union' (European Commission, 2015). The CEF includes one branch of funding for electricity and gas infrastructure projects, with €5.85 billion available for 'projects of common interest' between 2014 and 2020. The CEF follows on from the previous TEN-E Decisions with Regulation (EU) No. 347/2013 on guidelines for trans-European energy infrastructure. It supports pipeline connections to import natural gas, interconnections within the EU and LNG terminals. As such, the CEF does not deviate from previous policy directions. Discussions around the 'Energy Union', however, place climate change rhetoric far more at the centre, with 'an ambitious climate policy at its core' (European Commission, 2015: 2). Nevertheless, despite recognising the need to move the EU's economy away from reliance on fossil fuels, the Commission still places security of supplies of natural gas as the second of 15 points summarising the Energy Union, highlighting the potential of LNG and of the Southern gas corridor in responding to worries of gas supply interruptions in the future (p. 19). It may still be true to say that the long-term implications of the EU's decarbonisation objective have yet to be fully integrated into policy on natural gas infrastructure (Dupont and Oberthür, 2015a). As the EU moves towards reducing its GHG emissions by 40 per cent by 2030 and by 80 to 95 per cent by 2050, the move away from fossil fuels, including natural gas, still needs to take place at a sufficient pace.

References

Bahgat, G. (2006). Europe's Energy Security: Challenges and Opportunities. *International Affairs*, 82(5), 961–975.

Baran, Z. (2007). EU Energy Security: Time to End Russian Leverage. *The Washington Quarterly*, 30(4), 131–144.

BP. (2014). *BP Statistical Review of World Energy June 2014*. London: British Petroleum.

Brown, S. P. A. and Huntington, H. G. (2008). Energy Security and Climate Change Protection: Complementarity or Tradeoff? *Energy Policy*, 36, 3510–3513.

Council of the European Union. (2005a). Common Position Adopted by the Council on 8 December 2005 with a View to the Adoption of a Decision of the European

Parliament and of the Council Laying Down Guidelines for Trans-European Energy Networks and Repealing Decisions No. 96/391/EC and No. 1229/2003/EC. 10720/1/05.
Council of the European Union. (2005b). Statement of the Council's Reasons. Common Position Adopted by the Council on 8 December 2005 with a View to the Adoption of a Decision of the European Parliament and of the Council Laying Down Guidelines for Trans-European Energy Networks and Repealing Decisions No. 96/391/EC and No. 1229/2003/EC. 10720/1/05.
Dupont, C. (2015). Decarbonization and EU Relations with the Caspian Sea Region. In C. Dupont and S. Oberthür (eds), *Decarbonization in the European Union: Internal Policies and External Strategies* (pp. 180–200). Houndmills: Palgrave Macmillan.
Dupont, C. and Oberthür, S. (2015a). Conclusions: Lessons Learned. In C. Dupont and S. Oberthür (eds), *Decarbonization in the European Union: Internal Policies and External Strategies* (pp. 244–265). Houndmills: Palgrave Macmillan.
Dupont, C. and Oberthür, S. (eds). (2015b). *Decarbonization in the European Union: Internal Policies and External Strategies*. London: Palgrave Macmillan.
ECF. (2010). *Roadmap 2050. A Practical Guide to a Prosperous, Low Carbon Europe.* Brussels: European Climate Foundation.
EREC. (2010). *Re-thinking 2050: a 100% Renewable Energy Vision for the European Union.* Brussels: European Renewable Energy Council.
Eurogas. (2011). *Eurogas Roadmap 2050*. Brussels: Eurogas.
European Commission. (1995). Green Paper: for a European Union Energy Policy. COM(94) 659.
European Commission. (2000). Green Paper: Towards a European Strategy for the Security of Energy Supply. COM(2000) 769.
European Commission. (2001). Proposal for a Decision of the European Parliament and of the Council Amending Decision No 1254/96/EC Laying Down a Series of Guidelines for Trans-European Energy Networks. COM(2001) 775.
European Commission. (2003a). Commission Staff Working Paper. Decision of the European Parliament and of the Council Laying Down Guidelines for Trans-European Energy Networks and Repealing Decisions No. 96/391/EC and No. 1229/2003/EC. Extended Impact Assessment. SEC(2003) 1369.
European Commission. (2003b). Proposal for a Decision of the European Parliament and of the Council Laying Down Guidelines for Trans-European Energy Networks and of a Decision of the European Parliament and of the Council Laying Down Guidelines for Trans-European Energy Networks and Repealing Decisions No. 96/391/EC and No. 1229/2003/EC. COM(2003) 742.
European Commission. (2005). Communication from the Commission to the European Parliament Pursuant to the Second Subparagraph of Article 251 (2) of the EC Treaty Concerning the Common Position of the Council on the Adoption of Decision of the European Parliament and of the Council Laying Down Guidelines for Trans-European Energy Networks and Repealing Decisions No. 96/391/EC and No. 1229/2003/EC. COM(2005) 716.
European Commission. (2006). Opinion of the Commission pursuant to Article 251 (2), Third Subparagraph, Point (c) of the EC Treaty, on the European Parliament's Amendments to the Council's Common Position Regarding the Proposal for a Decision of the European Parliament and of the Council Laying Down Guidelines for Trans-European Energy Networks and Repealing Decisions No. 96/391/EC and No. 1229/2003/EC. COM(2006) 381.

European Commission. (2008a). Communication on the Directive 2004/67/EC of 26 April 2004 Concerning Measures to Safeguard Security of Natural Gas Supply. COM(2008) 769.
European Commission. (2008b). Second Strategic Energy Review: an EU Energy Security and Solidarity Action Plan. COM(2008) 781.
European Commission. (2009a). Proposal for a Regulation of the European Parliament and of the Council Concerning Measures to Safeguard Security of Gas Supply and Repealing Directive 2004/67/EC. COM(2009) 363.
European Commission. (2009b). Proposal for a Regulation of the European Parliament and of the Council Establishing a Programme to Aid Economic Recovery by Granting Community Financial Assistance to Projects in the Field of Energy. COM(2009) 35.
European Commission. (2010a). Communication from the Commission: Energy Infrastructure Priorities for 2020 and Beyond – a Blueprint for an Integrated European Energy Network. COM(2010) 677.
European Commission. (2010b). *Country File: Libya*. Brussels: Market Observatory for Energy.
European Commission. (2010c). *Country File: Russia*. Brussels: Market Observatory for Energy.
European Commission. (2010d). *Country File: Ukraine*. Brussels: Market Observatory for Energy.
European Commission. (2010e). Energy Infrastructure Priorities for 2020 and Beyond – a Blueprint for an Integrated European Energy Network. COM(2010) 677.
European Commission. (2010f). Report from the Commission for the Council and the European Parliament on the Implementation of the European Energy Programme for Recovery. COM(2010) 191.
European Commission. (2011a). Communication from the Commission: Energy Roadmap 2050. COM(2011) 885/2.
European Commission. (2011b). *Country File: Algeria*. Brussels: Market Observatory for Energy.
European Commission. (2011c). *Country File: Norway*. Brussels: Market Observatory for Energy.
European Commission. (2011d). Impact Assessment Accompanying the Document: Energy Roadmap 2050. SEC(2011) 1565 Part Two.
European Commission. (2012). *Country File: Turkey*. Brussels: Market Observatory for Energy.
European Commission. (2014). European Energy Security Strategy. COM(2014) 330.
European Commission. (2015). Energy Union Package: a Framework Strategy for a Resilient Energy Union with Forward-Looking Climate Change Policy. COM(2015) 80.
European Council. (2009). *Presidency Conclusions*, March 2009. Brussels: Council of the European Union.
European Parliament. (2005). Report on the Proposal for a Decision of the European Parliament and of the Council Laying Down Guidelines for Trans-European Energy Networks and Repealing Decisions No. 96/391/EC and No. 1229/2003/EC. Committee on Industry, Research and Energy. A6-0134/2005.
European Parliament. (2009). *Directorate-General for Internal Policies. Policy Department A: Economic and Scientific Policy. 'Gas and Oil Pipelines in Europe'*. Brussels: European Parliament.
Eurostat. (2015). Eurostat Online Database. Available at: ec.europa.eu/eurostat/data/database.

Gas LNG Europe. (2014). *LNG Map*. Brussels: Gas Infrastructure Europe.

Giordano, V., Gangale, F., Fulli, G. and Sánchez Jiménez, M. (2011). *Smart Grid Projects in Europe: Lessons Learned and Current Developments*. Brussels: Joint Research Centre.

Hauser, H. (2011). European Union Lobbying Post Lisbon: an Economic Analysis. *Berkeley Journal of International Law*, 29(2), 680–709.

Heaps, C., Erickson, P., Kartha, S. and Kemp-Benedict, E. (2009). *Europe's Share of the Climate Challenge: Domestic Actions and International Obligations to Protect the Planet*. Stockholm: Stockholm Environment Institute.

IEA. (2011). *World Energy Outlook 2011: Are We Entering a Golden Age of Gas?* Paris: OECD/International Energy Agency.

IEA. (2012). *Natural Gas Information*. Paris: OECD/International Energy Agency.

IEA. (2014). *Energy Policies of IEA Countries: European Union, 2014 Review*. Paris: OECD/International Energy Agency.

Larsson, R. L. (2007). *Nord Stream, Sweden and Baltic Sea Security*. Stockholm: FOI – Defence Research Agency.

Reichardt, K., Pfluger, B., Schleich, J. and Marth, H. (2012). With or Without CCS? Decarbonising the EU Power Sector. *Responses Policy Update*, 3(July 2012).

Sattich, T. (2015). Electricity Grids: No Decarbonization without Infrastructure. In C. Dupont and S. Oberthür (eds), *Decarbonization in the European Union: Internal Policies and External Strategies* (pp. 70–91). Houndmills: Palgrave Macmillan.

Shell International BV. (2008). *Shell Energy Scenarios to 2050*. The Hague: Shell.

Skea, J., Chaudry, M. and Wang, X. (2012). The Role of Gas Infrastructure in Promoting UK Energy Security. *Energy Policy*, 43, 202–213.

Stephenson, E., Doukas, A. and Shaw, K. (2012). Greenwashing Gas: Might a 'Transition Fuel' Label Legitimize Carbon-Intensive Natural Gas Development? *Energy Policy*, 46, 452–459.

Vasileiadou, E. and Tuinstra, W. (2013). Stakeholder Consultations: Mainstreaming Climate Policy in the Energy Directorate? *Environmental Politics*, 22(3), 475–495.

Von Hirschhausen, C. (2010). Developing a Supergrid. In B. Moselle, J. Padilla and R. Schmalensee (eds), *Harnessing Renewable Energy in Electric Power Systems: Theory, Practice, Policy* (pp. 181–206). London: Earthscan.

6 Explaining climate policy integration

Policy, politics, context and process

In the previous three chapters, I described the evolution of a number of EU energy policies between 2000 and 2010 and measured the extent to which long-term climate policy objectives were integrated into their process and output. It is clear from the discussion in those chapters that, even in the best cases, CPI is insufficient when considered from the long-term perspective of achieving GHG emissions reductions of 80 to 95 per cent by 2050. Considering the 2050 goal was agreed by EU member states and politicians themselves and is simply a new formulation of the longstanding goal to limit global temperature increase to 2° Celsius, the question to be asked is why has CPI remained insufficient?

In this chapter, I apply the four explanatory variables described in Chapter 2 to the story of CPI in each of the policies discussed in the previous chapters to try to understand why such levels of CPI were found in each case. The four explanatory variables are the *nature of functional interrelations*; the *extent of political commitment*; the *institutional and policy context* (each of which can explain the level of CPI in the policy output and policy process); and the *process dimension* (which can explain the level of CPI in the policy output only). I do not discuss separately the effect of each variable on the level of CPI in the policy process and the policy output, as output can to some extent be assumed to flow from process (Briassoulis, 2005a).

Results and variation: insufficient climate policy integration

Chapters 3 to 5 examined the levels of CPI in three cases of EU energy policy over the decade 2000 to 2010. Despite all these policies falling within the realm of 'energy policy', different levels of CPI were found. Furthermore, the levels of CPI were generally lower than initially expected. Even policies such as renewable energy policies and energy efficiency of buildings policies – which can be defined also as climate policies – did not demonstrate high levels of CPI. Table 6.1 summarises the levels of CPI found in the policy process and output of the cases.

While the result that the levels of CPI vary among the chosen case studies is not necessarily surprising, the fact that even the best cases for integrating climate policy objectives did not achieve sufficient or high levels of CPI is surprising.

Table 6.1 Summary of levels of CPI in the policy process and output of the case studies

Case	CPI in the policy process	CPI in the policy output
2001 RES-E Directive	Low to medium	Low
2009 RE Directive	Medium to high	Low to medium
2002 EPBD	Low to medium	Low
2010 EPBD recast	Low to medium	Low
2003, 2006 TEN-E guidelines	Very low	Very low
2009 EEPR Regulation	Very low	Very low

Combating climate change was one of the main motivations behind EU decisions to advance policy on renewable energy and energy efficiency in the 2000s. Yet, these policy measures do not meet expectations of requirements to achieve the goal of limiting global warming to 2° Celsius or of reducing GHG emissions in the EU by 80 to 95 per cent by 2050. A further surprise in the results is the little evolution in the level of CPI in the policy measures over the course of the decade under examination. Except for some increase in the level of CPI in policy process and output of RE policy between 2001 and 2009, there were no measurable improvements in the level of CPI in energy performance of buildings policy and natural gas import infrastructure policy. The shortening of the timeframe for achieving policy objectives (to 2050) is one likely explanation for the lack of evolution in CPI. Finally, the level of CPI in the policy process may not be so clearly linked to the level of CPI in the policy output as one could expect from an assumption that the output follows on from the process. This may be related to the indicators used for measuring CPI in the policy process. As noted in a number of the chapters, even where pro-climate voices were missing from the policy process that did not mean that climate arguments were also missing. In some cases, the policymakers at the forefront of the policy process may already have been convinced of the importance of developing policy to combat climate change. As such, some of the measurements described here for the level of CPI in the policy process may be considered as artificially low. A different indicator, perhaps linked to a discourse analysis of the actors involved, may provide a higher level of CPI for the process in some cases. Nevertheless, the indicator describing whether or not functional interrelations with long-term climate policy objectives are recognised provides an assessment of whether policymakers had internalised the need to promote CPI. Furthermore, even with higher levels of CPI in the policy process, the levels of CPI in the policy output would remain at the low levels described in Table 6.1. It may then be possible to hypothesise that there is more of a disconnect between the policy process and output than could be assumed, if such results were found.

In summary, we can highlight a number of results from the study of the cases. First, CPI within a specific policy field (namely, energy) varies across policy measures. This may be indicative also of a lack of integration across policy files within the energy sector, with different policies emphasising one or two of the EU's energy policy objectives (sustainability, competitiveness and security of supply) and little interaction among even energy policymakers (interviews 9, 12, 25 and 26). Second, CPI is insufficient in these cases to help achieve long-term climate policy objectives. Measured against the benchmark of achieving EU climate policy objectives to 2050, the policy outputs do not close the gap between the status quo and the 2050 goal in a timely enough manner. Third, there is little change in the levels of CPI in the cases over time, possibly due to the shortening deadlines towards 2050 (even as specific policy measures become more stringent). CPI in energy policy could thus be described as 'too little, too late' and that the EU is in fact engaged in something like 'catch-up governance' (Dupont and Oberthür, 2015a), where improvements in past policy measures do little to close the gap to 2050. Searching for explanations for these findings forms the central focus of this chapter.

In the next sections, I draw upon the explanatory framework described in Chapter 2 to explain these results. This involves discussing how *functional interrelations*; *political commitment*; and the *institutional and policy context* can explain the level of CPI found in the policy process and output of the cases discussed in Chapters 3–5. Additionally, I examine the role of the *policy process* as an explanatory variable for the level of CPI in the policy output only.

Four variables to explain CPI: policy, politics, context and process

In Chapter 2, I described an explanatory framework inspired from literature on policy integration and theories of European integration that included four variables with the potential to explain the levels of CPI found in the cases. These four variables are wrapped up in the broader notions of 'policy', 'politics', 'context' and 'process'. The first variable under discussion in this section is the nature and recognition of functional policy interrelations. This particular variable is embedded within understandings of policy and process, how policy is made and how general problems become policy problems (Kingdon, 2003). The second variable, political commitment, describes the politics surrounding the policy development and output at the time negotiations were ongoing. The third variable, the institutional and policy context, links to the politics of a particular policymaking moment, and the broader context within which the policy is made. Finally, the policy process itself can help explain the levels of CPI in the policy output, on the assumption that the policy output flows from the policy process in a policy cycle (Howlett and Ramesh, 2003).

The nature and recognition of functional interrelations

In this section, I discuss the nature of the functional interrelations between the three policy sectors discussed in Chapters 3–5 and long-term climate policy objectives. Where functional interrelations exist, some demand for CPI may occur. Functional interrelations may be more or less direct or indirect (or closer or farther apart); and more or less synergistic or conflictual. The more direct and synergistic the functional interrelations, the more likely that higher levels of CPI can be expected, as policy connections are harmonious and obvious. The opposite can also be expected to hold true.

First, the functional interrelations between RE policy and climate policy are direct and synergistic. The objectives of the 2001 RES-E Directive were to increase the share of RE in the EU to 12 per cent by 2010, corresponding to a 22 per cent increase in RE in electricity. For the 2009 RE Directive, the objective is to increase the share of RE in final EU energy consumption to 20 per cent by 2020. These objectives directly affect the long-term climate policy objectives in a synergistic way. In other words, increasing the share of (most sources of) RE reduces the level of GHG emissions by displacing fossil fuels. Scenarios to achieve long-term climate policy objectives of decreasing GHG emissions in the EU by between 80 and 95 per cent point to the highly important role to be played by RE in the future (see, for example, EREC, 2010; European Commission, 2011; Heaps, Erickson, Kartha and Kemp-Benedict, 2009; WWF, 2011). Although the long-term climate policy target was not described in the same terms at the time of the negotiation of the 2001 RES-E Directive, the level of ambition required was nonetheless known. Limiting global temperature increase to 2° Celsius – commonly agreed, since the 1990s, as the long-term climate policy target – requires clearly ambitious policy measures to reduce the emissions of GHGs. Developing RE policy contributes to achieving these goals.

Policymakers in the EU linked the development of RE policy to the achievement of climate policy goals, with the achievement of climate objectives being part of the main motivations behind the policy measures. As discussed in Chapter 3, however, this recognition did not necessarily extend to the long-term climate policy goals. The nature of the functional interrelations certainly facilitated the integration of climate policy goals in the case of RE policy, but other factors may have prevented full integration of long-term goals.

Second, the nature of the functional interrelations between the EPBDs' objectives and long-term climate policy objectives is considered direct and synergistic. In other words, achieving the aims of the 2002 and 2010 EPBDs (reducing energy consumption of buildings) can help achieve the objectives of climate policy to reduce GHG emissions (by displacing demand for fossil energy). As with the RE sector, scenario and roadmaps towards decarbonisation by 2050 highlight the importance of energy efficiency measures.

The nature of these functional interrelations was also recognised by policymakers, as discussed in Chapter 4. For the 2002 EPBD, the recognition of the synergistic nature of the functional interrelations with climate policy objectives provided one motivation for policy development – along with concerns over energy supplies. The long-term climate policy objectives were not necessarily central to policy development at this time. In the case of the 2010 EPBD recast, synergies were especially emphasised for the benefit of achieving climate objectives. Not only were policy measures in the EPBD about reducing energy consumption in buildings, but there were also measures included to promote the use of renewable energy for any remaining energy consumption of nearly zero energy new buildings (Arts. 6, 9), meaning that GHG emissions in new buildings, at least, ought to be eliminated under EPBD provisions. The limited nature of some of the provisions in the EPBD (such as the focus on new buildings), however, demonstrates also that policymakers did not necessarily take the long-term climate policy objectives seriously into consideration in policy development.

Third, the functional interrelations between external gas infrastructure support and long-term climate policy objectives to reduce GHG emissions in the EU by 2050 are indirect and conflictual. Where both TEN-E and the EEPR policy measures seek to support, financially and politically, the development of external gas infrastructure in the EU, these measures can only be considered as indirectly related to (in other words, they are not closely linked) and in conflict with long-term climate policy objectives. The continued use of natural gas without proven CCS technology in the long term threatens the achievement of long-term climate policy objectives (gas becomes part of the climate problem). Policies to support the expansion of external gas infrastructure risk locking in fossil fuel infrastructure. This in turn may lead to continued gas consumption, even when alternatives are becoming available. These functional interrelations are indirect, in that the construction of external gas infrastructure in itself does not necessarily lead to increases in GHG emissions. It is the risk of carbon lock-in that is at stake here. Increased capacity in external gas infrastructure means, at best, that investments are diverted away from more climate-friendly energy projects (such as renewable energy projects, energy efficiency improvements, and so on). At worst, it means the EU's energy system is locked-in to a fossil fuel infrastructure that stunts moves towards decarbonisation (Dupont and Oberthür, 2012; Dupont, 2015). The functional interrelations are conflictual because achieving the policy objectives of supporting further natural gas importing infrastructure is counter to the long-term climate policy objective of reducing GHG emissions by reducing fossil fuel consumption.

Chapter 5 outlined how the functional interrelations with long-term climate policy objectives were not recognised by policymakers in this sector. Since the functional interrelations between the policies are indirect, they may thus be obscured or hidden in the policy process. Long-term climate policy objectives may not have been recognised, therefore, also as short-term benefits of switching from coal to gas, and the short-term role of gas in combating climate change, are highlighted (European Parliament, 2005). When functional interrelations are indirect (and thus unclear) and conflictual, it may be more difficult for stakeholders and policymakers to see (and then articulate) the interrelations between the policy objectives. Furthermore, in the case of natural gas infrastructure policy, very few internal or external climate voices were present to push the long-term policy objectives onto the table – where policymakers failed to recognise the interrelations with long-term climate policy, there were no other voices highlighting these interrelations to them.

The nature of functional interrelations leads to more or less favourable contexts for CPI. The more direct and synergistic the functional interrelations,

Table 6.2 The nature of functional interrelations and the expected effect on levels of CPI

Case	Nature of functional interrelations	Effect on CPI
RE policy	Direct and synergistic	Most favourable
EPB policy	Direct and synergistic	Most favourable
External gas infrastructure policy	Indirect and conflictual	Unfavourable

the more likely these will be taken up during the policy process. The more synergistic or harmonious policy objectives, the easier that both policy objectives can be served through policy development. The more direct the interrelations between policies, the more likely that policymakers will recognise the opportunities for policy integration. Thus, the nature of functional interrelations between climate policy and RE and EPBD policy can be considered *most favourable* to the advancement of CPI. The interrelations between climate policy and gas import infrastructure policies, on the other hand, can be considered *unfavourable* for CPI (see Table 6.2).

Political commitment

When discussing the explanatory role of political commitment to levels of CPI, I assess, first, the political commitment of the EU and its institutions to combating climate change in an overarching sense. Second, I assess levels of political commitment to promoting CPI in the cases.

First, as climate change rose on the international political agenda throughout the 1990s and 2000s, the EU began to pay more attention and aimed to become an international leader on combating the problem (Oberthür and Roche Kelly, 2008; Wurzel and Connelly, 2011a). However, the EU had difficulty in bringing its rhetoric on the international stage in line with its internal policies. In the early 2000s, the EU finally made headway on agreeing internal climate policies, such as the Emissions Trading System, and policies that aimed to respond to the climate challenge, such as in the RE and energy efficiency realms. This helped demonstrate that the EU was serious about taking action on climate change, although the policies of the early 2000s were not yet ambitious enough to make a significant impact. Internal policy development was linked very clearly to the international negotiations on climate change, with the policy measures adopted in the early 2000s being part of the EU's response to its commitments under the Kyoto Protocol to reduce GHG emissions by 8 per cent between 2008 and 2012, compared to 1990 levels (European Commission, 2000a). This was a period in which the EU expressed its desire to provide international leadership on climate change. The early 2000s was a period of symbolic, rather than concrete or credible, EU political commitment to combating climate change generally, which can be measured as *medium* levels of political commitment.

The level of political commitment of the EU to combating climate change evolved over the course of the 2000s. From about 2007 onwards, the EU

158 *Explaining climate policy integration*

demonstrated particularly high levels of political commitment to the climate issue generally. The European Council (the highest political level of the EU) came out in March 2007 in favour of a binding target to reduce GHG emissions by 20 per cent by 2020, compared to 1990 levels, combined with a binding target to increase the share of RE in the EU to 20 per cent by 2020 and a non-binding target to improve energy efficiency by 20 per cent by 2020 (European Council, 2007). This period marks a shift in EU commitment to leadership on climate change from symbolic leadership based on rhetorical statements, to credible leadership with policy measures to move forward on combating climate change in the EU (Eckersley, 2012; Wurzel and Connelly, 2011b). The EU was particularly anxious at this time to demonstrate its credible leadership in advance of the 2009 international negotiations at the fifteenth Conference of the Parties (COP) to the UNFCCC in Copenhagen, and pushed for the adoption of domestic policy measures in advance of this meeting. At this point in time, political commitment to combating climate change generally was at a *high* level in the EU, with statements expressing the political commitment followed up with concrete policy measures. For the early and late 2000s, the overarching levels of political commitment to combating climate change can provide some context to explain the levels of CPI found in the cases (see Table 6.3).

Second, I assess political commitment to *climate policy integration* in the cases. With regard to the 2001 RES-E Directive, the Commission justified its proposal with reference to the Kyoto Protocol commitment, and clearly saw RES-E policy mostly as a climate policy measure, although other motivations were also present (European Commission, 2000b: 2–3). Additionally, the Commission highlighted that the RES-E Directive helps achieve EU objectives to integrate environmental protection requirements into other policies but these motivations did not push the Commission to propose a strong, ambitious Directive with mandatory targets. The Parliament's proposed amendments aimed to strengthen the Commission's proposal, especially by making targets mandatory (European Parliament, 2000). In the Council, the level of political commitment to CPI in the 2001 RES-E Directive is different to both the Commission and the Parliament. In the lead up to the Commission's proposal, the Council pushed the Commission to propose legislation on RE (ENDS Europe, 1999), but during the negotiations on the proposal, several member states lowered the overall ambition and reinforced the idea of indicative non-binding targets (Council of the European Union, 2001). In this case, while the Commission and the Parliament can be considered as demonstrating *medium to high* and *high* levels of political commitment, the Council lowers the overall level of political commitment to integrating climate policy objectives into the RES-E Directive with its *low to medium* levels of political commitment. Thus, EU political commitment to CPI in the 2001 RES-E Directive is summarised as *medium*.

In the later RE policy developments, it is clear that the proposal for an RE Directive in 2008 was considered by EU policymakers as a direct response to commitments to combat climate change. The Commission stated in the proposal that: 'the challenges of climate change caused by anthropogenic emissions of greenhouse gases, mainly from use of fossil energy, need to be tackled effectively and urgently' (European Commission, 2008: 2). The RE Directive is an energy policy that aims to achieve climate policy objectives. The same level of political commitment to advancing CPI can be found in the Parliament. The Parliament called for more ambitious policy on RE and pushed for interim targets to achieve the 2020 target, including sanctions for states that did not achieve these. In the Council, there was also quite a considerable level of political commitment to advancing CPI in RE policy at this time. This is probably linked to the upcoming international climate negotiations in Copenhagen in 2009. The Presidencies in 2008 (Slovenia and France) were particularly motivated to achieve agreement before the end of the year (Boasson and Wettestad, 2013). While member states were uninterested in Parliament's proposal for binding interim targets and sanctions, they did not negotiate on the binding 20 per cent target and they also agreed with Parliament that sustainability criteria would be required for biofuels (ENDS Europe, 2008b). In the 2009 RE case, it seems that all three deciding institutions in the EU demonstrated *high* levels of political commitment to advancing CPI in RE policy.

In the 2002 EPBD, the Commission framed its proposal as responding to the EU's commitments under the Kyoto Protocol to reduce GHG emissions in the EU by 8 per cent by 2008 to 2012 compared to 1990 levels. Additionally, the Commission included a Recital in its proposal for the EPBD specifically referring to the legal objective of the EU to integrate environmental concerns into other sectoral policies. This remained throughout the negotiations and is the first Recital of the final Directive. However, the motivation to respond to climate change is balanced against other motivations in the Commission's proposal such as security of supply. The Parliament similarly was interested in achieving co-benefits through the EPBD for competitiveness, climate change and security of supply. Its main amendments to the Commission's proposal were related to improving considerations on costs and financing for the EPBD, rather than strengthening its outputs for the benefit of the climate. For the Council, there were questions about the EU being the appropriate level for policy on the energy performance of buildings. The Council attempted to water down the original proposal by introducing more flexibility, and managed to negotiate a longer time-frame for the implementation of the Directive – effectively postponing any remaining benefits that could be accrued through the policy. In this case, where both the Parliament and the Commission could be said to demonstrate *medium* levels of commitment to CPI, the Council is rather on the *low* end of the scale. Some added weight could be provided to the role of the Council in this case, given the little ability of Commission and Parliament to change its views, so an overall score of *low to medium* seems most justified.

With respect to political commitment to CPI in the 2010 EPBD recast, this is tempered by the lower level of political commitment to advancing measures on energy efficiency generally. Of the 2020 targets agreed by the European Council in 2007, only the target to save 20 per cent of energy by 2020 compared to BAU was non-binding. The Commission's proposal for a recast of the 2002 EPBD did

160 *Explaining climate policy integration*

not go much beyond the original Directive. The proposal was sharply criticised as not being ambitious enough by several stakeholders (interviews 5, 6, 7 and 10). The Parliament took up much of the slack and went on to provide its 'best first reading ever' from a climate perspective, according to one NGO representative (interview 5). However, the Parliament soon lost momentum as it entered trialogue negotiations and compromised heavily with the Council. The Council, having showed little enthusiasm for binding energy efficiency targets, continued along a path of little commitment to advancing CPI in the EPBD recast. Member state concerns over flexibility, financial costs and considerations of national differences meant they had little appetite for pushing for ambitious EU-level policy for the EPBD recast. The one exception to this low level commitment may have been Sweden, which held the Presidency in the second half of 2009 in the lead up to the Copenhagen climate negotiations. However, in this case, it seems that Sweden was rather more interested in reaching agreement before the end of its Presidency term than in ensuring ambitious policy measures for combating climate change (Boasson and Wettestad, 2013: 146; interview 6). There is therefore a somewhat mixed story for the political commitment to CPI in the 2010 EPBD case. Where the Commission can be said to demonstrate *medium* levels of CPI, the Parliament may demonstrate *medium to high* levels (high at the beginning of the process, but compromising quickly with Council showing reduced levels of commitment). The Council again demonstrates *low* levels of commitment to CPI, providing an overall *medium* level of political commitment. Again, in this case, it should be noted that the role of the Council seems to have trumped the roles of the Commission and Parliament.

For the 2003 and 2006 TEN-E Decisions, there is little or no evidence of political commitment to pushing CPI. This is perhaps unsurprising, given the lack of the recognition of the functional interrelations between long climate policy objectives and the objectives of external gas infrastructure support policy. No link to CPI is made in the political statements or the subsequent policy output in the TEN-E Decisions. The very suggestion of labelling external gas infrastructure projects as having 'priority' and as being 'projects of European interest' suggests that the political commitment in this case rather supports the expansion and continuation of policies for developing external gas infrastructure (contrary to what could be considered supporting CPI, which would be reducing and eliminating such policies). When it comes to external gas infrastructure policies, the overarching concern is energy security and climate concerns rarely enter the picture. In this case, there is little difference among the EU institutions, for whom energy security is the overarching and dominating motivation, drowning out concerns for climate change. In the Parliament, for example, energy security has often been weighted by the majority of ITRE committee members as the most important energy policy objective, above competitiveness and sustainability, reducing opportunities for commitment to the promotion of CPI (interviews 25 and 26). In the Council, neither climate change nor promoting CPI seemed to enter discussions on TEN-E. In summary, for the 2003 and 2006 TEN-E Decisions, there is *no* or at best *low* levels of political commitment to advancing CPI.

Explaining climate policy integration 161

Table 6.3 Overarching political commitment to combating climate change and to advancing CPI in the cases

Case	Overarching political commitment	Political commitment to CPI
2001 RES-E Directive	Medium	Medium
2009 RE Directive	High	High
2002 EPBD	Medium	Low to medium
2010 EPBD recast	High	Medium
2003, 2006 TEN-E guidelines	Medium	None/low
2009 EEPR Regulation	High	None/low

With respect to political commitment to CPI in the external gas infrastructure policy of the 2009 EEPR Regulation, there is again little or no evidence of commitment to CPI. Despite the overarching context of high commitment to combating climate change at this time, it was rather the economic crisis and the ensuing poor investment context for large infrastructure projects that motivated the EEPR Regulation. With regard to external gas infrastructure in particular, the 2006 and 2009 gas crises helped place energy security at the forefront of considerations on infrastructure projects. The Southern gas corridor, for example, received promises of €300 million for two proposed pipelines, with the major aim of sourcing natural gas from the Caspian region (and reducing dependence on Russia). In discussions on this and other projects, there is *no* or *low* commitment to the advancement of CPI in the three EU institutions.

For a summary of the extent of political commitment to combating climate change and to advancing CPI in each of the three cases, see Table 6.3.

Institutional and policy context

There are several internal and external institutional and policy contextual factors that may also provide some explanation for the levels of CPI found in the policy processes and outputs of the cases. These are linked to the internal institutional set-up within the EU, the policy objectives being pursued at the time, and the international context of geopolitics, energy politics and climate negotiations.

First, the institutional context at the time of negotiations may shed some light on why such levels of CPI were found. In each of the three cases, the Directives, Decisions and Regulations were adopted under the co-decision procedure (today known as the ordinary legislative procedure). This means that QMV was the decision rule in the Council. In theory, such a voting rule prevents a single state vetoing any decision and could help promote CPI (if a majority of states were in favour of strong climate policy objectives). In practice, however, Council seems to prefer to agree policies through consensus building (Novak, 2010; Tsebelis, 2013). This means that the potential for QMV to advance CPI is not always realised and individual member state concerns about a certain policy proposal are worked out in the policy negotiations. Thus, it cannot be said that the decision-making rules provided *favourable* or *less favourable* circumstances for advancing CPI. This context is rather *neutral* towards advancing CPI, therefore.

Consultation and impact assessment procedures regarding policy proposals developed over time in the EU. Where consultation procedures are open and transparent, we can expect more opportunities for internal and external pro-climate stakeholders to raise their concerns in the policy process. The impact assessment procedure allows internal and external stakeholders to scrutinise the considerations taken into account during the development of a policy proposal. In the cases examined in Chapters 3–5, three examples stand out where consultation procedures were less than open and transparent. The 2001 RES-E Directive and the 2002 EPBD were proposed before consultation procedures were formally adopted in the Commission. This meant that there were few official procedures for external stakeholders to get involved in the early stages of the policy process. These standards were adopted as part of the later better regulation push in 2002 (European Commission, 2002). The third example is the 2009 EEPR Regulation. Due to the expressed urgency of adopting measures to release funds in favour of infrastructure projects in the context of the economic and financial crises, no impact assessment and no open consultation took place for this particular policy file. The speed of the negotiation process (less than six months from proposal to adoption) on the EEPR Regulation meant that in-depth negotiations among the policymakers did not really take place, and opportunities for informal lobbying were limited. Except for in these particular cases, consultation and impact assessment procedures were in place. Thus, this reality may lead to a *more favourable* or *neutral* context when the procedures were in place, and *less favourable* when such procedures did not exist.

During the course of the 2000s, the EU also underwent two rounds of enlargement in 2004 and 2007, increasing the number of member states negotiating on a particular file to 27. In theory, reaching a compromise among so many member states was expected to pose difficulties for agreeing far-reaching and ambitious policy, while a new influx of MEPs to the Parliament was also thought to threaten EU environmental governance. Literature has found that enlargement has slowed down decision-making in the EU and resulted in Parliament proposing less radical amendments to environmental proposals (Burns, Carter and Worsfold, 2012; Hertz and Leuffen, 2011; Thomson, 2009). In RE policy, enlargement did not seem to have a great effect on the final policy output. In the 2009 RE Directive negotiations, Poland finally agreed to the negotiated compromise as it gained certain concessions in parallel negotiations on the revision of the ETS (Boasson and Wettestad, 2013). Here we can see a new member state is making use of package negotiations to push an agenda. The overall impact on the RE Directive was limited, though. In the EPBD cases, it was not necessarily the new member states that resulted in the policy being watered down. Even member states that had already implemented energy performance standards domestically (such as Denmark) did not push for ambitious measures at the EU level (Boasson and Wettestad, 2013). The Parliament did propose amendments to strengthen the EPBD proposal, but did not succeed in pushing these amendments through. However, in the 2010 recast the new member states did pose a challenge when it came to policy measures referring to the existing building stock – as buildings in many Central and Eastern European countries are particularly poor on energy performance standards. With no concrete measures to renovate buildings, for the EPBD, we can say that enlargement may have resulted in a *less favourable* context for advancing CPI. In the external gas infrastructure case, it is enlargement that is the reason for negotiating a revision of the TEN-E guidelines so soon after the 2003 Decision was adopted. Many of the new member states are particularly reliant on Russia for supplies of natural gas, although their levels of gas consumption may not be as high as in some other member states. They were generally in support of any policy measure that improved their energy security and helped diversify sources of gas away from reliance on Russia. In this case, enlargement was likely *less favourable* to advancing CPI in external gas infrastructure policy, as more member states were in favour of promoting such infrastructure without regard to long-term climate policy objectives.

The three cases also show a certain amount of path dependency over the course of their development in the 2000s. Path dependency is an element of historical institutionalist perspectives in particular that highlights the importance of history. It can describe why certain policies or institutions remain stable over time (Peters, Pierre and King, 2005; Pollack, 2009). While there was a certain amount of change in the cases examined in Chapters 3–5, with targets becoming binding in RE policy, with measures becoming slightly more stringent in the EPB Directives, and with more money being allocated to gas infrastructure projects in the TEN-E and EEPR cases, the main lines of policy action stayed rather similar in each case. It is in the RE case that policy change is most evident, as the RE Directive moved from one of indicative targets for electricity to binding targets for overall energy consumption, divided among member states. Here, path dependency provided perhaps a *neutral* context. In the EPBDs, the lessons of the failure of the 2002 EPBD to reduce the energy consumption of buildings significantly did not lead to radical policy change proposals. The 2010 EPBD simply reformulated some of the main measures of the 2002 EPBD, but still without binding targets. The 2009 EEPR Regulation followed the same path blazed for years by the TEN-E guidelines, by assigning financial and political support to certain infrastructure projects that are labelled of 'European interest' or as 'priority projects'. The amounts of financial assistance assigned in the 2009 EEPR Regulation were considerably higher than the previous TEN-E Decisions, but otherwise the EEPR did not deviate from the TEN-E policy measures. For these two cases, then, path dependency may have provided *less favourable* context for advancing CPI. As combating climate change becomes more urgent over time, these particular policy measures are less likely to evolve to integrate climate policy objectives more than they had previously.

For each case, there were particular internal institutional contexts that also might have played a role in whether or not CPI could be significantly advanced.

164 *Explaining climate policy integration*

In the 2009 RE Directive, the proposal was made as part of a package of policy measures on climate and energy. As such, the fate of the RE Directive was very much tied to the advancement of the rest of the files (on ETS, CCS and an Effort Sharing Decision). This meant that a certain amount of trading of positions among the files occurred, such as Poland agreeing on the RE Directive due to concessions accorded in the ETS negotiations. However, this sort of trading was relatively limited in the later stages of the negotiations on each of the files and many of the contentious issues on the RE Directive were agreed fairly early on in the process (ENDS Europe, 2008c; ENDS Europe, 2008d). On the entire climate and energy package, the European Council and Presidency played unusual roles to facilitate agreement. The European Council announced political agreement on all files of the package before the final vote took place in the Parliament. The 2010 EPBD recast then was proposed and negotiated around the same time as the climate and energy package, although negotiations on the EPBD recast took longer than for the package. The EPBD involved many of the same policymakers as the negotiations on the climate and energy package (Boasson and Wettestad, 2013). This package negotiation may have provided a *favourable* context for advancing CPI in the RE and EPBD cases. The proposal and negotiations on the 2009 EEPR Regulation were particularly quick and took place within the context of European Parliament campaigning during an election year. This quick process during such a time meant that many Parliamentarians were not even aware of the policy file at this time, and some ITRE committee members did not recall having discussed the file at all (interviews 25 and 26). Such a context meant that the Parliament may not have had the opportunity to consider the proposal in depth and is likely to have been *less favourable* for advancing CPI.

Second, there are several external policy context factors that may also shed light on the levels of CPI found in the cases. The international climate negotiation schedule is important in especially the RE and EPBD cases. In 1997, the EU committed to reduce its GHG emissions by 8 per cent between 2008 and 2012, compared to 1990 levels. The 2001 RES-E Directive and the 2002 EPBD were both intended as instruments that could contribute to achieving this objective. In addition, they also served to show the international community the EU's leadership and commitment to combating climate change (Oberthür and Roche Kelly, 2008). These same ambitions were not part of the TEN-E Decisions. Later policy agreements on the 2009 RE Directive and the 2010 EPBD recast were negotiated and adopted with a certain haste, which was due to the upcoming 2009 Conference of the Parties (COP) to the UNFCCC in Copenhagen in December. The 2009 RE Directive was adopted earlier in 2009 in advance of the COP, and informal agreement on the 2010 EPBD was also achieved in time for the Copenhagen negotiations. Thus, in the case of RE and EPB policy, the EU's ambitions to be a leader in the international negotiations on climate change are generally *favourable* to advancing CPI in domestic policy measures in the EU. This favourable context must come with some caveats, however. It is possible that the time pressure to reach agreement on the 2009 RE Directive and the

Explaining climate policy integration 165

2010 EBPD might have led to early compromise on certain elements of the policy measures (interview 6). Furthermore, the international climate negotiations agenda did not affect the level of CPI in external natural gas infrastructure policy. Policy measures that are less directly linked to reducing GHG emissions, therefore, may have remained untouched by the international context.

Towards the end of 2008, Europe entered a period of financial and economic crisis. This dual crisis impacted the policy measures adopted. The negotiations on the RE Directive were completed shortly after the crisis began. A clause allowing the Directive to be reviewed was inserted towards the end of the negotiations, perhaps as a result of the crises (ENDS Europe, 2008a). Many of the sticking points in the negotiations on the 2010 EPBD recast were linked to financing measures and costs of the policy. More optimistic voices tried to highlight the job creating benefit of ambitious EPBD policy measures (interviews 5 and 10), but finally no specific financing targets were agreed. In external natural gas infrastructure, the crises were the main motivations for the proposal, speedy negotiation and adoption of the 2009 EEPR Regulation. As described in Chapter 5, the very existence of policy measures to promote external natural gas infrastructure is counter to long-term climate policy objectives. Thus, the overall effect of the economic and financial context at the end of the 2000s is *less favourable* towards advancing CPI.

Broader energy security concerns may also have played a role in the development of CPI in these policies. The 2006 and 2009 gas crises, in which Russia temporarily cut supplies of natural gas to Ukraine, affecting some EU member states downstream, raised the issue of energy security on the agenda. Where this might have benefited CPI was in RE and EPB policy. Advancing the share of RE and reducing energy consumption of buildings in the EU can be seen as measures that provide co-benefits for energy security and the environment. Energy security was indeed a stated motivation in these cases, but, especially for the EPBD recast, this did not push CPI further. Conversely, energy security concerns hampered the advancement of CPI in the policy on external natural gas infrastructure. The energy security concerns so trumped considerations of long-term climate policy objectives that new infrastructure projects were provided financial and political backing despite 2050 goals that require considerable reductions in fossil fuel consumption in the EU. For the three cases, therefore, we cannot say that energy security concerns were helpful in pushing CPI, but provided a *less favourable* context for the external gas sector, and a *neutral* context for RE and EPB policies.

Table 6.4 provides an aggregate result of the potential for the internal institutional set-up in the EU and the external policy context to advance CPI in each of the cases discussed in Chapters 3–5. As the table shows, the internal and external contexts vary in terms of how favourable they are towards CPI advancement, both over time and among the policy measures. What the table does not account for, however, is that some contexts may be more important than others in a particular policy field. For example, the external context of perceived vulnerability to energy shortages is clearly a major motivation for policy on

166 *Explaining climate policy integration*

Table 6.4 The potential of the internal institutional context and external policy context to advance CPI in the cases

Case	Internal institutional context	External policy context
2001 RES-E Directive	Less favourable	More favourable
2009 RE Directive	More favourable	Neutral
2002 EPBD	Neutral	More favourable
2010 EPBD recast	More favourable	Neutral
2003, 2006 TEN-E guidelines	Less favourable	Less favourable
2009 EEPR Regulation	Less favourable	Less favourable

external gas infrastructure, and thus a barrier towards CPI. In RE policy, the international climate negotiations agenda, especially towards the end of the 2000s, may have played a more significant role in pushing CPI than the economic crisis did in preventing CPI. Taking an aggregate view, therefore, provides a certain understanding of the internal and external contexts but may not allow for a more nuanced understanding of what particular context was most significant for CPI in each case.

Process dimension

Finally, I discuss how CPI in the policy process itself can be deployed as an explanatory variable for understanding CPI in the policy output in particular. The policy output is the final policy agreement and it is preceded by negotiations and discussions among policymakers and stakeholders. Thus, it can be assumed that there are links between the level of CPI in the policy process and the output and that the policy output is unlikely to come from thin air.

In the policy process leading to the 2001 RES-E Directive, the level of CPI was found to be *low to medium* (see Table 6.1). CPI in the policy output was found to be *low*. In the policy process leading to the agreement on the 2009 RE Directive, the level of CPI was found to be *medium to high*, while the policy output was *low to medium*. These results point to a certain correlation between the policy process and output. The level of CPI in the policy output is not expected to exceed the level found in the process, and that is the case for RE policy (Briassoulis, 2005b). Beyond this, however, we need to break down the result for the level of CPI in the policy process to understand in more detail where the lower levels of CPI in the policy output may have come from. For the 2001 RES-E Directive, the *low* levels of involvement of external pro-climate stakeholders stand out. There was little involvement of external pro-climate stakeholders beyond informal lobbying activities, as there were no open consultation procedures and no impact assessments in place at this time. Furthermore, the recognition of the functional interrelations focused predominantly on the short-term climate policy objectives of achieving the EU's commitments under the Kyoto Protocol. Perhaps the low involvement of external pro-climate stakeholders reinforced the lack of recognition of long-term climate objectives. Poor access and no procedures for the involvement of external pro-climate stakeholders meant they could not further highlight the functional interrelations between ambitious RE policy and the long-term climate policy objectives. In the case of the 2009 RE Directive, breaking down the level of CPI in the policy process shows that for the Council, the interaction of internal and external pro-climate stakeholders was *low*. In fact, the Council may have been the main reticent actor in the policy process, but the *low* result of its openness to pro-climate stakeholders was bolstered in the measurement of CPI in the policy process by the involvement of these actors in the Commission and Parliament. Thus, it is possible that the Council played a key role in weakening the level of CPI in the final policy output below a level that could have been expected based on the level of CPI in the policy process overall.

In the 2002 and 2010 EPBDs, the level of CPI in the policy process was *low to medium* and did not improve over the course of the decade. Similarly, the level of CPI in the policy output of both Directives was *low*. Nevertheless, in breaking down the level of CPI in the policy process to the specific indicators, we see that there was a higher involvement of external pro-climate stakeholders in the lead up to the 2010 recast Directive than in the 2002 EPBD. However, this change was insufficient to improve the overall levels of CPI in the policy process. Policymakers initially recognised the functional interrelations between EPBD policy measures and climate policy objectives (especially for the short-term climate policy objectives under the Kyoto Protocol), but little attention was given in both cases to long-term climate policy objectives throughout the process. In both the 2002 and the 2010 EPBD, it seems that the Council played a crucial role in lowering the levels of CPI. The member states were the most reticent internal actors for agreeing policy measures, and the Council was the least open institution for receiving input from external pro-climate stakeholders (interviews 5 and 6). The arguments of subsidiarity put forward and the lack of either enthusiasm or interest of the Council seemed to outweigh other factors in the policy process and could plausibly thus be considered a main element of the process dimension that resulted in lower levels of CPI in the policy output (Boasson and Dupont, 2015).

In the case of the external gas infrastructure policy in the TEN-E revisions and the 2009 EEPR Regulation, the level of CPI in both the policy process and policy output was measured (see Table 6.1) as *very low*. The level of CPI in the policy process is expected to affect the level in the policy output – without evidence of CPI in the policy process, it would be surprising to suddenly find high levels of CPI in the policy output. While each of the three indicators in the measurement of CPI in the policy process play a role in helping to explain CPI in the policy output, the lack of the recognition of functional interrelations between external gas infrastructure policy and long-term climate policy objectives seems crucial in this case. In theory, environmental and pro-climate stakeholders could have seized on opportunities to involve themselves in the policy process, with online consultation procedures and informal lobbying activities in the lead up to the adoption of the TEN-E Decisions (although the formal option was not available in the 2009 EEPR Regulation process). However, even when one pro-climate internal stakeholder was involved to an extent (the ENVI committee in the Parliament during the negotiations on the 2006 TEN-E revisions), they did not express concern that the external gas infrastructure support would be in conflict with long-term climate policy objectives. In this respect, even the pro-climate stakeholders did not recognise (or articulate or act upon) the functional interrelations with long-term climate policy objectives.

In summary, the level of CPI in the policy process may indeed play an explanatory role for understanding the level of CPI in the policy output. In each case, CPI in the policy process was either higher than or the same as the level of CPI in the policy output. However, real explanatory value comes when the policy process result is broken down into separate indicators. Therefore, for the RE case, we note that the low involvement of stakeholders in early policy developments and the recognition of short-term functional interrelations (rather than interrelations with long-term climate goals) may provide more nuanced explanation. In EPB policy, it seems that the Council was the most important actor in the policy process – both in terms of its own negotiation position and in terms of its lack of openness to external stakeholders. In external gas infrastructure, the lack of recognition of functional interrelations may have prevented any advancement of CPI. This lack of recognition was on the part of policymakers and stakeholders.

Bringing the explanatory variables together

Through the above discussion, it is clear that certain variables played a greater role in certain cases, meaning that the four variables did not carry equal explanatory weight. Each of the four explanatory variables played more or less crucial roles in the explanation of the levels of CPI found. From this analysis, it may be possible to identify a hierarchy of importance among the explanatory variables for an analysis of CPI. In addition, some of the variables proved most valuable as explanatory variables only in combination with others or in support of other variables. Table 6.5 provides an overview of the contributions of the explanatory variables to understanding the levels of CPI in the cases.

Further empirical research may be required to test the explanatory framework and develop a larger evidence base from which to draw conclusions about the hierarchy or constellations of explanatory variables that prove most crucial for a study on CPI. However, the discussion above can provide some first clues about how the explanatory variables interact with each other.

First, the nature and recognition of the functional interrelations is a crucial first-order explanatory variable for understanding CPI. Second, political commitment plays a role in the explanation of the levels of CPI once the functional interrelations have been recognised and articulated. Unless there is some

Table 6.5 Summary of the explanatory variables for levels of CPI in the case studies

Case	Nature of functional interrelations	Political commitment to climate change	Political commitment to CPI	Institutional context	Policy context	Process dimension
2001 RES-E Directive	Direct and synergistic	Medium	Medium	Less favourable	More favourable	Low guaranteed involvement of pro-climate stakeholders
2009 RE Directive	Direct and synergistic	High	High	More favourable	Neutral	Key role of Council
2002 EPBD	Direct and synergistic	Medium	Low to medium	Neutral	More favourable	Key role of Council
2010 EPBD recast	Direct and synergistic	High	Medium	More favourable	Neutral	Key role of Council
2003, 2006 TEN-E Decisions	Indirect and conflictual	Medium	None or low	Less favourable	Less favourable	Lack of recognition of functional interrelations
2009 EEPR Regulation	Indirect and conflictual	High	None or low	Less favourable	Less favourable	Lack of recognition of functional interrelations

recognition of the functional interrelations by the policymakers in a policy sector, political commitment to combating climate change or to advancing CPI is unlikely to play a role in advancing CPI. Third, the institutional and policy context provides a background context that can add further explanatory nuance, when the nature and recognition of the functional interrelations and the level of political commitment do not seem to provide sufficient explanation. Fourth, the process dimension is most useful as a complementary explanatory variable when broken down into its component elements. Therefore, rather than a hierarchy of variables, we could perhaps describe the variables as interrelating with each other in a loosely sequential manner, and as reinforcing the overall explanation.

In RE policy, the two most crucial explanatory factors for the insufficient levels of CPI in the policy process and output are the increasing levels of political commitment over time, and the role of pro-climate stakeholders. Pro-climate stakeholders had easier access to the policy process in the negotiations on the 2009 Directive than in the 2001 RES-E Directive. In the case of RE policy, those policymakers involved had already understood the nature of the functional interrelations with climate policy, although the long-term objectives were not always central to the negotiations. Heightened political commitment, especially from the Council in the run up to the Copenhagen negotiations thus comes as a clearly important explanatory variable, since the issue of functional interrelations is understood.

In EPB policy, the level of CPI in the policy process and output did not change over the course of the first decade of the twenty-first century. Although the policy measures did undergo some change, there was little advancement in CPI in particular, considering the closing timeframe to achieve the 2050 climate goals. In the case of the two EPBDs, functional interrelations were already recognised to some extent by policymakers. The synergistic nature of these relations could in theory have helped push for more advanced levels of CPI than was found. But considering that the functional interrelations were recognised, that the general context was relatively favourable to advancing CPI and that political commitment to combating climate change and advancing CPI existed to a certain extent, we can rather suppose that the key role played by the Council is the crucial variable to explain the lack of change over time. Indeed, the EU member states seem to have been the major blocking force for strengthening policy to advance CPI (Boasson and Dupont, 2015).

Finally, in external natural gas infrastructure policy, the first step in advancing CPI has been missed. The functional interrelations between policies that support natural gas infrastructure and policies to combat climate change are conflictual and rather indirect. This may result in policymakers (and stakeholders) neglecting to recognise that the policies do indeed interact. When breaking down the process dimension, this seems to be the reason behind the lack of evidence of CPI in both the process and output. Where policymakers do not even link the policy to climate objectives, there can be no expectation for CPI. Therefore, even with medium or high general levels of political commitment to combating climate change, without recognition of functional interrelations, such political commitment does not trickle down to policy development.

The discussion on the role of functional interrelations and their recognition as a first-order explanatory variable for CPI also points to a further finding. While recognising that a certain policy interrelates with climate policy objectives is a necessary first step to advancing CPI, it is also important to underline the long-term nature of the climate policy objective (to 2050). It seems that even in those cases (RE and EPB policy) where interrelations were recognised, these focused rather on short-term policy objectives and not the long-term goals. Had the long-term climate policy objectives been seriously considered in the policy development of the three cases, more ambitious policy outputs could have been expected (Dupont and Oberthür, 2015b). Instead, the insufficiency of the levels of CPI for the 2050 goal in each case is striking, even when policymakers recognised the nature of the functional interrelations.

In addition, by breaking down the policy process, a further finding seems to point to the role of pro-climate stakeholders. Where pro-climate stakeholders were a part of the policy process and negotiations, they could not push policymakers to agree ambitious policy measures that are in line with the 2050 goal. Furthermore, pro-climate stakeholders cannot always be guaranteed to highlight the functional interrelations with long-term climate policy objectives, even if the opportunities for them to be involved in the policy process are present. This is particularly evident in the TEN-E Decisions, where pro-climate stakeholders were mostly absent from the policy process, although usual consultation procedures were present. In the negotiations on the EEPR, such stakeholders were hardly involved in even informal lobbying. It may be the case that EU institutions are not yet sufficiently 'open' in the policy process to allow such voices to be raised, but it is also possible that EU policymakers cannot rely on stakeholders to bring the climate message into debates.

Summary

In sum, we can say that the explanatory variables do indeed help explain the levels of CPI in the policy process and output of the cases examined in Chapters 3–5. The main findings from the empirical research include that: first, CPI within a specific policy field (namely, energy) varies across policy measures. Second, CPI is insufficient in the three cases to achieve 2050 climate policy objectives. Third, there is little change in the levels of CPI in the cases over time, possibly due to the shortening deadlines towards 2050 (even as specific policy measures become more stringent) and the EU engaging in 'catch-up governance'. Fourth, from the application of the explanatory framework, it is also found that the recognition of the nature of the functional interrelations between the policy being developed and long-term climate policy objectives is a first-step variable for the advancement of CPI. Fifth, procedures in EU policymaking may be insufficient to ensure that pro-climate stakeholders recognise these interrelations, enter the policy discussions and raise pro-climate arguments.

In terms of how the four explanatory variables interact with each other, each variable plays a certain role in the context of the cases studied. The role of the recognition of functional interrelations is shown to be a crucial first step for any advancement of CPI. From there, the remaining variables interact in different ways. Political commitment proved important for advancing CPI, but where the political commitment to advancing CPI is not high in the Council (as with the EPB example), then CPI may be lower than expected. This leads to recognition of the key role played by the Council in several cases. Whether the Council is in favour or against strong policy measures, it seems to be the EU institution that holds the most weight in the decision process. The role of the Council may be linked to the institutional and external policy contexts, but the relationships are not easy to assess. For further nuanced assessment of the role of the explanatory variables in analysing and understanding CPI, more empirical research would be required to take the number of cases beyond those few examined here.

References

Boasson, E. L. and Dupont, C. (2015). Buildings: Good Intentions Unfulfilled. In C. Dupont and S. Oberthür (eds), *Decarbonization in the European Union: Internal Policies and External Strategies* (pp. 137–158). Houndmills: Palgrave Macmillan.

Boasson, E. L. and Wettestad, J. (2013). *EU Climate Policy: Industry, Policy Innovation and External Environment*. Farnham: Ashgate.

Briassoulis, H. (2005a). Analysis of Policy Integration: Conceptual and Methodological Considerations. In H. Briassoulis (ed.), *Policy Integration for Complex Environmental Problems: the Example of Mediterranean Desertification* (pp. 50–80). Aldershot: Ashgate.

Briassoulis, H. (ed.) (2005b). *Policy Integration for Complex Environmental Problems: the Example of Mediterranean Desertification*. Aldershot: Ashgate.

Burns, C., Carter, N. and Worsfold, N. (2012). Enlargement and the Environment: The Changing Behaviour of the European Parliament. *Journal of Common Market Studies*, 50(1), 54–70.

Council of the European Union. (2001). Common Position Adopted by the Council on 23 March 2001 with a View to the Adoption of Directive of the European Parliament and of the Council on the Promotion of Electricity Produced from Renewable Energy Sources in the Internal Electricity Market. 5583/1/01.

Dupont, C. (2015). Decarbonization and EU Relations with the Caspian Sea Region. In C. Dupont and S. Oberthür (eds), *Decarbonization in the European Union: Internal Policies and External Strategies* (pp. 180–200). Houndmills: Palgrave Macmillan.

Dupont, C. and Oberthür, S. (2012). Insufficient Climate Policy Integration in EU Energy Policy: the Importance of the Long-Term Perspective. *Journal of Contemporary European Research*, 8(2), 228–247.

Dupont, C. and Oberthür, S. (2015a). Conclusions: Lessons Learned. In C. Dupont and S. Oberthür (eds), *Decarbonization in the European Union: Internal Policies and External Strategies* (pp. 244–265). Houndmills: Palgrave Macmillan.

Dupont, C. and Oberthür, S. (eds). (2015b). *Decarbonization in the European Union: Internal Policies and External Strategies*. Houndmills: Palgrave Macmillan.

Eckersley, R. (2012). Does Leadership Make a Difference in International Climate Politics? Paper presented at the *International Studies Association Annual Conference*. San Diego, CA.

ENDS Europe. (1999). EU Ministers Request Renewables Framework, *12 May 1999*.

ENDS Europe. (2008a). Crisis Leaves the Climate Glass Half Empty, *19 December 2008*.

ENDS Europe. (2008b). EU Renewable Energy Law Deal Confirmed, *9 December 2008*.

ENDS Europe. (2008c). EU States Reach Accord on Renewables Plan, *29 October 2008*.

ENDS Europe. (2008d). New EU Renewable Energy Law 'Finalised'. *04 December 2008*.

EREC. (2010). *RE-thinking 2050: a 100% Renewable Energy Vision for the European Union*. Brussels: European Renewable Energy Council.

European Commission. (2000a). Communication from the Commission on EU Policies and Measures to Reduce Greenhouse Gas Emissions: Towards a European Climate Change Programme (ECCP). COM(2000) 88.

European Commission. (2000b). Proposal for a Directive of the European Parliament and of the Council on the Promotion of Electricity from Renewable Energy Sources in the Internal Electricity Market. COM(2000) 279.

European Commission. (2002). Towards a Reinforced Culture of Consultation and Dialogue – General Principles and Minimum Standards for Consultation of Interested Parties by the Commission. COM(2002) 704.

European Commission. (2008). Proposal for a Directive of the European Parliament and of the Council on the Promotion of the Use of Energy from Renewable Sources. COM(2008) 19.

European Commission. (2011). Communication from the Commission: Energy Roadmap 2050. COM(2011) 885/2.

European Council. (2007). *Presidency Conclusions*, March 2007. Brussels: Council of the European Union.
European Parliament. (2000). Report on the Proposal for a European Parliament and Council Directive on the Promotion of Electricity from Renewable Sources in the Internal Electricity Market. A5-0320/2000.
European Parliament. (2005). Report on the Proposal for a Decision of the European Parliament and of the Council Laying Down Guidelines for Trans-European Energy Networks and Repealing Decisions No 96/391/EC and No 1229/2003/EC. A6-0134/2005.
Heaps, C., Erickson, P., Kartha, S. and Kemp-Benedict, E. (2009). *Europe's Share of the Climate Challenge: Domestic Actions and International Obligations to Protect the Planet*. Stockholm: Stockholm Environment Institute.
Hertz, R. and Leuffen, D. (2011). Too Big to Run? Analysing the Impact of Enlargement on the Speed of EU Decision-Making. *European Union Politics*, 12(2), 193–215.
Howlett, M. and Ramesh, M. (2003). *Studying Public Policy: Policy Cycles and Policy Subsystems*, 2nd edn. Oxford: Oxford University Press.
Kingdon, J. W. (2003). *Agendas, Alternatives, and Public Policies*, 2nd edn. London: Longman.
Novak, S. (2010). Decision Rules, Social Norms and the Expression of Disagreement: the Case of Qualified-Majority Voting in the Council of the European Union. *Social Science Information*, 49(1), 83–97.
Oberthür, S. and Roche Kelly, C. (2008). EU Leadership in International Climate Policy: Achievements and Challenges. *International Spectator*, 43(3), 35–50.
Peters, B. G., Pierre, J. and King, D. S. (2005). The Politics of Path Dependency: Political Conflict in Historical Institutionalism. *Journal of Politics*, 67(4), 1275–1300.
Pollack, M. A. (2009). The New Institutionalisms and European Integration. In A. Wiener and T. Diez (eds), *European Integration Theory*, 2nd edn. (pp. 125–143). Oxford: Oxford University Press.
Thomson, R. (2009). Actor Alignments in the European Union Before and After Enlargement. *European Journal of Political Research*, 48(6), 756–781.
Tsebelis, G. (2013). Bridging Qualified Majority and Unanimity Decision-making in the EU. *Journal of European Public Policy*, 20(8), 1083–1103.
Wurzel, R. K. W. and Connelly, J. (2011a). Introduction: European Union Political Leadership in International Climate Change Politics. In R. K. W. Wurzel and J. Connelly (eds), *The European Union as a Leader in International Climate Change Politics* (pp. 3–20). London: Routledge.
Wurzel, R. K. W. and Connelly, J. (eds) (2011b). *The European Union as a Leader in International Climate Change Politics*. London: Routledge.
WWF. (2011). *The Energy Report: 100% Renewable Energy by 2050*. Gland, Switzerland: World Wide Fund for Nature.

7 Conclusions

This concluding chapter provides an overview of the main findings and draws out implications from the research that contribute to broader discussions of environmental and climate policy integration and to debates about the future of climate and energy policy in the EU.

Review of research question and findings

The research question described in Chapter 1 formed the basis for the ensuing discussions and empirical case research. The question this book sought to answer was: 'What is the extent of climate policy integration into the EU's energy policy, and why?' Answering this question required, first, measuring the extent of CPI in EU energy policy, and second, applying an explanatory framework with a number of variables to establish why such a level of CPI exists. The question therefore led to a research design that focused in-depth on a number of cases of EU energy policy. By examining the evolution of the levels of CPI in these cases over time, it also became possible to pinpoint whether changes or differences in the explanatory variables could account for the levels of CPI found.

There were a number of background puzzles that inspired this research question, including the role of the energy sector in causing climate change, the adoption of an 'integrated' climate and energy package of policy measures at the EU level in 2009 and the EU's ambitions for leadership on climate change at the international level. These developments aroused curiosity about the reality of CPI in the EU's energy sector. Some initial guiding questions included: just how integrated are the EU's climate and energy policies? Has the level of CPI changed? What can explain potential variation in the level of CPI in EU energy policies?

Academic work on the specific issue of CPI is slowly growing, but studies that focus on the EU level are few (Adelle and Russel, 2013; Knudsen, 2012). There are several studies analysing CPI at national level (see, for example, Kivimaa and Mickwitz, 2006, 2009; Mickwitz et al., 2009), and much conceptual work on the concept of environmental policy integration (EPI) that provides a natural home for further conceptual development of CPI (Herodes, Adelle and Pallemaerts, 2007; Jordan and Lenschow, 2010; Lenschow, 2002; Persson, 2007). Discussions of CPI (often considered synonymous to 'climate mainstreaming') among EU policymakers have been occurring, especially in the context of adapting to the effects of climate change (European Commission, 2009, 2013). In some respects, CPI has gained more traction with policymakers than EPI ever did (Adelle and Russel, 2013).

To respond to the research question and to examine the reasons behind the levels of CPI found in EU energy policy, I chose to follow the development of three cases of EU energy policy over the time period 2000 to 2010. Each of these three policy areas differs with regard to the objectives, measures and actions in place. Renewable energy policy involves *increasing domestic supply* of new sustainable energy sources. Energy performance of buildings policy targets the demand side of the energy sector, and aims to *reduce energy consumption* in the EU. Policies to promote importing natural gas infrastructure deal with *external energy relations* and aim to ensure stable supplies of imported gas. In each of the cases discussed in Chapters 3–5, policy developments took place in the early and late 2000s, so that process tracing over time was possible.

The case studies were selected based on expected variation on the dependent variable – the extent of CPI (Gerring, 2008). After an initial survey of the potential cases, I expected variation in the results of CPI in the three cases, where the RE policy sector would demonstrate the highest levels of CPI, the EPB case would show medium levels, and the external gas case would show low levels of CPI. Each case study was analysed in-depth using *process tracing* techniques, with data drawn from *document analysis, literature review, media reports* and *interviews* (Barkin, 2008; Checkel, 2008; Hopf, 2004; Klotz, 2008).

Drawing on a conceptualisation of CPI that considers it to be a matter of degree (Bryner, 2012), I operationalised the concept by outlining a benchmark for *very high* levels of CPI for the policy process and the policy output. This very high level can correspond to suggestions that environmental or climate policy objectives should be assigned principled priority over the objectives of the policy sector into which they are being integrated (Lafferty and Hovden, 2003). When no weight is assigned to (in this case) climate objectives, Liberatore (1997) argues that there is a risk for policy dilution. Hence, the benchmark for *very high* levels of CPI in the policy output was developed with reference to the long-term objective of reducing GHG emissions in the EU by 80 to 95 per cent by 2050 (or ensuring temperature increase does not exceed 2° Celsius). For each case study, the question could then be asked (with reference to the many studies on the road to decarbonisation by 2050): what role will this policy sector play in achieving decarbonisation by 2050? If the policy output places that sector on a trajectory to achieving decarbonisation, it could then be considered to display evidence of *very high* levels of CPI. For the policy process, the *very high* benchmark is rather more qualitative. I assessed CPI in the process with reference to three factors, namely the involvement of pro-climate stakeholders internally to the EU policymaking process (such as DG Environment, the ENVI committee in the Parliament, the environment formation of the Council); the involvement of external pro-climate stakeholders in the policymaking process (such as environmental and climate NGOs, pro-climate policy industries and industry associations, including the RE and insulation industries); and the recognition of the functional interrelations between the policy being negotiated and long-term climate policy objectives. Each of these three factors was measured on a qualitative fivefold scale to arrive at an aggregated score for the level of CPI in the policy process, although for this project each indicator was assigned equal weight in the measurement of CPI in the policy process.

In response to the first part of the research question, different levels of CPI in EU energy policy between 2000 and 2010 were found. The assumptions that underlined the initial case study selection were not necessarily confirmed, however. Variation on the dependent variable (the extent of CPI) was observed, but not to the same levels as expected (see Table 6.1). Where *high* levels of CPI were expected in the RE case, the levels of CPI in 2001 turned out to be *low* in the policy output and *low to medium* in the policy process. In 2009, the level of CPI in the policy process had increased to *medium to high* but the level of CPI in the policy output had barely improved, measuring *low to medium*. Where *medium* levels of CPI were expected in the EPB case, the levels of CPI in the policy process of both the 2002 and 2010 Directives were actually *low to medium*. The levels of CPI in the policy output of both the 2002 and 2010 Directives were *low*. Finally, *low* levels of CPI were expected for the external gas infrastructure policy. In this case, expectations were relatively accurate; although the surprising story is that there is a *lack* of evidence of any CPI in this case. Levels of CPI in the policy output and process of the external gas infrastructure policy measures of the TEN-E guidelines were *very low*. These levels did not change in the EEPR Regulation in 2009, which also showed *very low* CPI in its policy process and policy output.

The next step in answering the research question required *explaining* the results found from the measurement of CPI in the policy process and output in each case. For this purpose I identified four explanatory variables: *functional interrelations; political commitment; institutional and policy context*, each of which could help explain the level of CPI in both the policy process and output; and *the process dimension*, which could help explain the level of CPI in the policy output only. In deriving these explanatory variables, I drew inspiration, first, from EPI literature and its many approaches and (often long) lists of variables, and, second, from general theories of European integration (Fioretos, 2011; Haas, 1958; Lafferty and Hovden, 2003; Moravcsik and Schimmelfennig, 2009; Persson, 2004; Pollack, 2009; Strøby-Jensen, 2007). Using these perspectives and theories to provide inspiration for the explanatory framework led to the development of a framework that is manageable and applicable to the EU level. I did not set out to test the validity of any of the individual theories, but rather drew from them to understand better the empirical reality of CPI in EU energy policy.

Each of these explanatory variables was discussed in Chapter 6 to explain the main results of the empirical research. These results were:

- CPI is insufficient in the three cases to achieve 2050 climate policy objectives;

- Levels of CPI vary across the cases (therefore CPI varies within the energy policy field);
- There was little change in the level of CPI in the three cases over the course of the decade 2000 to 2010.

The first result may not be surprising, given the difficulty for policymakers to integrate long-term considerations into day-to-day policymaking (Dupont and Oberthür, 2012; Hovi, Sprinz and Underdal, 2009; Voss, Smith and Grin, 2009). The very low levels of CPI actually found, however, are somewhat surprising, especially in cases where the policy can even be considered to respond directly to the climate problem (as with RE and EPB policies, for example). The second finding points to a lack of integration even *within* the energy policy sector in the EU. The three policy objectives of EU energy policy – sustainability, competitiveness and security – are not weighted equally across the various energy policies. Indeed, there seems to be little overlap and discussion among policymakers on different policy files (interviews 9 and 12), with policies emphasising one or two of the energy policy objectives. The weight applied to each of the three policy objectives also seems to shift over time, depending on external events, perceived threats and opportunities, and political alliances (interviews 25 and 26). Thus, we can say that even as climate policy objectives are insufficiently integrated into EU energy policy, so is EU energy policy insufficiently integrated internally. The third result, that the levels of CPI hardly improved over time, is linked to the shortening time horizon to 2050. As policy development aims to fix the failures of past policy efforts, the EU engages in 'catch-up governance' (Dupont and Oberthür, 2015). This is a pattern of policy development that follows a trend of incremental improvements on previous policies, rather than moving towards the 2050 goal with ambition and clarity. Hence, the EU agrees on policies that are 'too little, too late' for the achievement of climate policy objectives.

The discussion on the why such levels of CPI were found revealed two further findings:

- The recognition of the functional interrelations between policy objectives is a crucial first step to advancing CPI;
- Procedures in the EU that ensure climate stakeholders can enter the policy process may be insufficient to advance CPI.

These two findings are linked to the long-term nature of climate policy goals, and to the institutional set-up of policymaking in the EU. For the first finding, the crucial flaw in the recognition of functional interrelations in each of the cases was the lack of consideration of the long-term perspective. Where policies on RE and EPB were motivated (partly) by commitments to reduce GHG emissions, the policy measures put in place were not ambitious enough for achieving the sort of emission reductions required under the 2050 goal. Thus, the recognition of functional interrelations is a crucial first step for advancing CPI, but this recognition must include an understanding of the long-term implications of policy. The second finding may have implications for transparency and openness in EU policymaking. Although improvements in the ability of stakeholders to raise their voices and interests in policymaking was evident between 2000 and 2010, it should be noted that policymakers cannot necessarily rely on stakeholders to enter all relevant policymaking processes. The lack of climate stakeholders in the TEN-E and EEPR discussions is not necessarily due to a lack of consultation and informal lobbying opportunities, but may rather be due to climate stakeholders selecting the policy processes on which they would concentrate. Thus, the climate message was missing from debates on external natural gas infrastructure. Openness to internal and external climate stakeholders may thus be insufficient to ensure the advancement of CPI in the process and output of a particular policy measure.

In the next sections, I discuss the implications of these findings for scholarship and for policymaking.

Scholarly contributions

The focus of the research described in this book was not on theory testing, but rather on a qualitative empirical analysis questioning the accuracy of describing EU climate and energy policy as 'integrated'. The research, nonetheless, contributes to academic discussions on EPI and CPI, and also to literature analysing the EU's climate and energy policies.

The literature on EPI in the EU has a relatively long history, but there is continued disagreement on what EPI *is* and how it can be identified or advanced (Jordan and Lenschow, 2008, 2010; Nilsson and Persson, 2003; Nollkamper, 2002; Persson, 2004). CPI, in particular, is still a new topic in academic literature (Adelle and Russel, 2013; Ahmed, 2009; Rietig, 2012). This book contributes to the literature by distilling the conceptual frameworks developed in the extensive literature on EPI and CPI and, combined with inspiration from general theories of European integration, by deriving a manageable conceptual framework for analysis of CPI in the EU. First, I developed a conceptualisation of CPI that can capture differing levels of CPI in the policy process and output. Conceptualising CPI as a matter of degree allowed for an empirical investigation of the extent of CPI that could, methodologically, capture very high levels of CPI ('strong' CPI) and also very low levels of CPI ('weak' CPI) (Bryner, 2012; Jordan and Lenschow, 2008). I did not consider CPI as something that either exists or does not exist. A more nuanced conceptualisation and operationalisation allowed for the reality of different levels of CPI in EU energy policy to be revealed. Operationalising CPI in both the policy process and output is a further addition to the literature – where studies of CPI have taken place, they often focus on the policy output, while studies of EPI have most often been linked to procedures in the policy process (Adelle and Russel, 2013). This particular conceptualisation and operationalisation would benefit from further refinement through future empirical research, but already provides a significant contribution to advancing literature on both CPI and EPI.

Second, I proposed a framework for analysing CPI in the EU that is manageable, yet also aims for comprehensiveness. The analytical framework combines variables derived from the long-established EPI literature and several established theoretical perspectives on European integration (see Chapter 2). Although the explanatory strength of the variables differed depending on the empirical realities of the cases (see Chapter 6), it can certainly be said that the combination of these variables helped explain the empirical reality more fully than a reliance on an explanatory framework derived or inspired from a single theory or perspective. This framework can be tested in different policy sectors, and through its application, can be refined and improved, especially as regards the role of the process as an explanatory variable.

Third, the empirical analysis in this research project included an in-depth assessment of CPI in EU energy policy, and challenges the assumption that the 2009 climate and energy package of measures adopted in the EU represents evidence for the advancement of EPI or CPI. It is clear from the empirical evidence presented here that such assumptions are unfounded. Even the EU's RE policy, with the 2009 RE Directive adopted as part of the climate and energy package, displays at best *low to medium* levels of CPI in the policy output from 2000 to 2010. The empirical findings in this project should lead scholars of EPI/CPI to look twice before accepting claims of advanced policy integration in the EU. The empirical research presented here also points to the variation that can exist within what could be considered a coherent single policy field (namely energy policy).

Fourth, and beyond literature on EPI and CPI, therefore, this book also contributes to scholarly work on the development of climate and energy policy in the EU, and the links between EU internal policy development and external claims of 'leadership' on climate change (Oberthür and Roche Kelly, 2008; Torney, 2014; Wurzel and Connelly, 2011). The cases show how policymaking dynamics within the EU can vary according to the policy file, to the institutional and policy context around the time when the policy is being negotiated, but also due to the commitment of, particularly, member states to move forward on such policy measures. Within the EU, it has been understood that adding 'credibility' to claims of international leadership on climate change requires far-reaching and ambitious policy measures (Wurzel and Connelly, 2011). The contribution of the work in this book is to highlight that internal policies of the EU may prove more ambitious than those of its partners, but even these policies are insufficiently ambitious when considered against the scale of action required to 2050. Scholars could reflect on what such findings may mean for the EU's external climate change politics.

Fifth, the differences and changes in the levels of CPI in the *policy process* of some of the three case studies point to findings that are of interest to scholarly literature on interest representation in the EU (Eising and Lehringer, 2010; Gullberg, 2008a; Michaelowa, 1998; Tanasescu, 2009). In the RE case, it is particularly striking that the level of CPI in the policy process increased over time, but this did not result in a significant increase in the level of CPI in the policy output. For the EPB case, while more stakeholders were involved in negotiations on the 2010 EPBD than for the 2002 Directive, this did not result in an increase in the overall *low to medium* level of CPI in the policy process – possibly because of the particularly dominating role of the Council on this file. In the external gas infrastructure case, there was no evidence of CPI in the policy process, with external pro-climate stakeholders not involved, even though in the TEN-E Decisions, procedures were in place. These results led to reflection on the 'openness' of the EU energy policy processes to climate interest groups. In the late 2000s, EU institutions were generally more open to external stakeholders than in the early years of the same decade. Yet, this seems to be insufficient to ensure that climate voices enter the policy discussion. In other words, policymakers may not be able to assume that the interest and involvement of such stakeholders will be sufficient for pushing CPI, or that a general openness to stakeholders will ensure pro-climate stakeholders will be involved. Furthermore, as the EU institutions face more and more informal lobbying, climate voices may not be the strongest raised (Gullberg, 2008b; Hauser, 2011).

Sixth, the work presented in this book provides indications about the difficulties of the EU to respond effectively to long-term problems, such as climate change. The results show little change in the levels of CPI in the cases over time and provide further evidence for research on challenges for long-term policymaking in general (Dupont and Oberthür, 2012; Hovi et al., 2009; Voss et al., 2009). The limited improvement in the levels of CPI in the three cases studied points to the game of 'catch-up governance' that policymakers are engaged in: rather than agreeing sufficiently ambitious policy measures from the outset, policymakers are stuck in a cycle of trying to fix the failings of previous policy measures on a shortening deadline. Had the indicative target in the 2001 RES-E Directive been met, for example, more ambitious targets may have feasibly been adopted for the 2009 RE Directive, as the effort to increase the share of RE would already have begun. The same can be said of the EPB case, where the poor implementation of the 2002 EPBD meant that in the 2010 EPBD recast, policy measures were simply trying to catch-up on the ambitions set eight years previously. For both of these cases, the shifting BAU scenario in the later stages of the first decade of the twenty-first century gave less time for the policy measures agreed to demonstrate effective action on the road to decarbonisation. We could also argue that the findings on the lack of recognition of long-term functional interrelations hampered the ability of the EU to agree sufficiently ambitious policy from the start. Instead of integrating long-term climate policy objectives, the cases demonstrated striking levels of insufficient CPI. Further research about the ability of the EU, and other democratic systems, to overcome such problems when responding to long-term challenges would be welcome.

These are some of the main contributions to advancing scholarly work that can stem from the research presented in this book. For each body of literature, further research could supplement the initial findings described here.

What can stakeholders and policymakers do to promote CPI?

With climate policy integration, or 'climate mainstreaming', becoming a strategy in the EU to combat and adapt to climate change, the research presented here should raise interest among policymakers and climate stakeholders. Policymakers and stakeholders should consider the ramifications of the main result: that CPI is insufficient in the three EU energy policies examined here to achieve long-term (2050) climate policy objectives. What can be done to improve the record and ensure that climate goals to 2050 can be achieved?

Breaking patterns

The first challenge for stakeholders and policymakers involves breaking past patterns of incremental policy improvement. Catching up on past policy failures is an insufficient step forward when the timeline for achieving policy goals is shortening.

First, long-term planning for policy (towards 2050) has to be fully aligned with day-to-day politics and policymaking in the EU. This is a challenge especially for elected politicians working within an electoral cycle. Civil servants (often with the ability to remain longer in policymaking) may need to bear at least some of the burden of pushing to achieve these long-term objectives in practice. A consideration of long-term climate and decarbonisation policy objectives should be made an explicit part of the evaluation of any policy proposal. Where a policy sector objective is not in line with long-term objectives, policymakers may need to consider radical policy options (such as abandoning policies or making innovative adjustments to established policy frameworks). The Commission, in particular, can play the role of advancing discussions on long-term policymaking, for example, through green and white papers and by demonstrating long-term policymaking by example across its proposals. Stakeholders can also hold policymakers to account for any failures to integrate long-term climate policy objectives into policy development. The role of stakeholders to pressure, lobby and push for consideration of long-term goals may be critical to ensuring that any new procedures along these lines are respected across policy sectors. This may be a greater role than many environmental and climate NGOs can take on by themselves, so coalitions of stakeholders interested in achieving decarbonisation (including energy efficiency and renewable industries, human rights and poverty reduction NGOs, for example) may need to be established.

Second, especially when the nature of the functional interrelations between a policy sector and long-term climate policy objectives are indirect and/or conflictual, explicit efforts will be required to ensure the interrelations are recognised and considered in policymaking. These interrelations must first be recognised before policymakers can consider how to deal with them. A general openness to pro-climate stakeholders seems insufficient to ensure climate policy objectives are considered across policy sectors. This may require new procedural rules, such as an explicit requirement to consider and report on the potential functional interrelations with long-term climate policy objectives for every policy measure under development. The requirement in Article 11 of the TFEU to integrate environmental protection requirements into the definition and implementation of the Union's policies and activities is a legal requirement towards integration that has not been translated sufficiently into everyday policymaking. There may be a role for the Courts to pressure EU institutions to translate this integration requirement into stronger policy measures.

Setting priorities

Achieving decarbonisation may require some difficult decisions about priorities, but given the urgency of the climate problem (IPCC, 2013), with sustained political will, these decisions can be justified and taken (Dupont and Oberthür, 2015).

To meet the decarbonisation goal by 2050, trade-offs may be needed to prioritise the move to decarbonisation, even within energy policies. The three objectives of energy policy – security, competitiveness and sustainability – are not weighted evenly in policymaking and are often not even integrated with each other. To achieve decarbonisation, policies motivated by concerns for energy security, for example, may need to be adjusted in light of decarbonisation objectives. Continuing to secure supplies of fossil fuels into the future runs counter to decarbonisation objectives. Security objectives may need to be reframed in terms of decarbonisation-friendly policies (such as enhancing domestic supplies of renewable energy) (Casier, 2015). Furthermore, policies that aim to ensure low energy costs may need to be reassessed. If decarbonisation objectives require upfront capital investment, priority may need to be given to achieving this objective over the objective of ensuring costs are kept to a minimum. This particular trade-off is likely to prove controversial in times of economic recession. Innovative and creative linking of decarbonisation goals to objectives of job creation may be required, while supplementary social policies to aid the vulnerable may need to be negotiated. Additionally, assigning priority in policymaking to decarbonisation policies may need a high-level political decision (from the European Council), with more explicit commitment to this goal. A coalition of policymakers and politicians may be required to push for such an explicit commitment and decision.

Procedural measures

Finally, policymakers' buy-in to ensuring the integration of climate policy objectives within energy policy decisions is not yet obvious. Interaction among policymakers seems too limited, both formally and informally, to advance CPI (even among energy policymakers focused on different dossiers). Some procedures may need to be adopted to ensure that long-term climate policy objectives are considered across sectors, including obligations for closer collaboration among policymakers on the development of policy proposals. Requirements for long-term climate objectives to be considered in the impact assessment procedure and requirements for obligatory consultation with (internal and external) pro-climate stakeholders, regardless of the topic of the policy being developed may be needed. Such procedures may need to be developed to ensure climate change does not become further compartmentalised within certain committees or DGs (such as within the DG for Climate Action in the Commission). Similar procedures could be applied across the DGs of the Commission to ensure the interrelations of policies in, for example, research, transport, agriculture and so on with climate policy objectives become clear. Closer collaboration among committee members in the Parliament is also to be recommended, but cross-party group agreements on the necessity of combating climate change may be a better mechanism for ensuring long-term climate policy objectives are integrated into policy amendments and responses in the Parliament. Such cross-party agreement at national level within member states could lead to uploading this concern to the EU level in the Council. Procedures similar to those in the Commission could be also envisioned at the level of working groups in the Council, where policymakers receive training, awareness-raising and engage in exchange to ensure the climate policy objectives are kept to the fore.

Further issues, future research

While I set out to answer a seemingly simple question ('what is the extent of CPI in the EU's energy policy, and why?'), many other questions can be raised as a result of the analysis. Future research could focus on several empirical and conceptual questions.

The conceptual framework developed and applied in this book could benefit from wider application. Through testing and refining the framework, its strengths and weaknesses could be more clearly identified. One weakness of the framework identified through its application is that the explanatory role of the process requires the process dimension to be broken down to levels lower than the overall indicators before a valuable explanatory role may be revealed. Some weighting of the indicators in the process dimension may also be appropriate. Future research could help especially to refine this aspect of the framework by carrying out similar studies in different sectors of EU policymaking (such as transport, industry or agriculture policies). Additionally, the framework could benefit from slight adaptations and testing on other levels of policymaking – does the explanatory framework hold value also for analyses of CPI on the national level or international level? Such empirical research would enhance the evidence base for further theorising CPI in general.

In practice, CPI seems to be far more appealing to policymakers than EPI ever was (Adelle and Russel, 2013). Although this book reports that CPI in the EU's energy sector seems far from sufficient to achieve long-term climate policy objectives, there is nonetheless a certain level of CPI in some energy policies. The same may not be so easily identified for EPI. EPI may represent a higher level of integration that involves systemic and holistic thinking; it requires that the 'environment', broadly defined, be considered in all policymaking. By focusing on CPI, do academics and policymakers harm the overarching EPI notion? Will this focus lead to further erosion of holistic policymaking? Is climate change too politically salient an issue to allow other environmental considerations to enter the discussion? What consequences could a focus on CPI, rather than on EPI, actually mean for the environment? Many of these questions require philosophical theorising and cannot be answered solely with strong empirical analysis. Future research on CPI may thus require further thinking on these fundamental questions, which are unavoidably normative in nature. In some respects, these questions can be summed up in one: is CPI good for the environment? Some initial theorising on the links between EPI and CPI in the literature has begun, with Adelle and Russel (2013) discussing this specific issue. Taking this theorisation further and questioning the very justifications for CPI, combined with evidence-based empirical studies, may provide useful insights into how priorities should or could be assigned to different objectives in policymaking.

Finally, the issue of the challenges of long-term policymaking is one that recurred throughout this study. Climate change is a complex issue – GHGs emitted today will affect the future climate. With the IPCC's periodic assessment reports, the scientific consensus states that GHG emissions will have to peak as soon as possible before declining dramatically, on a global scale, in order to avoid potentially catastrophic consequences caused by climate change (IPCC, 2007, 2013). Such a global, complex problem, with decisions needed today to avoid future consequences, poses particular challenges for politicians and policymakers working within short electoral cycles. This 'wicked' problem, as some scholars name climate change (Jordan, Huitema, van Asselt, Rayner and Berkhout, 2010), requires a level of political courage and innovation that previously in history was mobilised in situations of serious conflict.

For climate policy development in the EU, the challenges of policymaking with objectives on a long-term horizon to 2050 are also present. Several interviewees mentioned that the 2050 objectives to reduce GHG emissions by 80 to 95 per cent are not considered in day-to-day policymaking. For politicians, the commitment is valid, but the measures to be put in place to reach the goal are not priorities (interviews 1, 4, 5, 6, 10, 23 and 26). In the case studies, a lack of consideration of the long-term perspective in policymaking was evident. Academic discussions on the challenges of solving climate change in democratic systems (Barker, 2008; Edmondson and Levy, 2013; Monaghan, 2013) could be complemented with further research on how the EU deals with the climate change issue, having committed itself to the long-term 2050 goal. Describing and analysing EU developments towards achieving long-term goals can be carried out using the 2050 goal as a benchmark, and/or by assessing CPI as carried out here. Linking this work to theories of democratic politics may prove fruitful for broadening the debate and discussion beyond the energy sector and even beyond the EU.

Finally, as the EU prepares its 2030 climate and energy framework and its plans for an Energy Union (European Commission, 2014, 2015), research on the evolution of the level of CPI in the energy sector could hold some promise. Perhaps the political commitment, institutional and policy context, recognition of functional interrelations and CPI in the policy process will evolve sufficiently in the negotiations and preparations of the new pieces of energy legislation to 2030. Research that considers taking the work presented in this book as a starting point for assessing CPI in the new energy policies of the EU could reveal further insights into how CPI has evolved and how it could be further promoted to achieve the 2050 decarbonisation goal.

References

Adelle, C. and Russel, D. (2013). Climate Policy Integration: a Case of Déjà Vu? *Environmental Policy and Governance*, 23(1), 1–12.

Ahmed, I. H. (2009). *Climate Policy Integration: Towards Operationalization*. DESA Working Paper No. 73: ST/ESA/DWP/73. New York: UN/DESA.

Barker, T. (2008). Climate Policy: Issues and Opportunities. In H. Compston and I. Bailey (eds), *Turning Down the Heat: the Politics of Climate Policy in Affluent Democracies* (pp. 15–32). Houndmills: Palgrave Macmillan.

Barkin, S. (2008). 'Qualitative' Methods? In A. Klotz and D. Prakesh (eds), *Qualitative Methods in International Relations: a Pluralist Guide* (pp. 211–220). New York: Palgrave Macmillan.

Bryner, G. C. (2012). *Integrating Climate, Energy and Air Pollution Policies* (with Robert J. Duffy). Cambridge, MA: MIT Press.

Casier, T. (2015). The Geopolitics of the EU's Decarbonization Strategy: A Bird's Eye Perspective. In C. Dupont and S. Oberthür (eds), *Decarbonization in the European Union: Internal Policies and External Strategies* (pp. 159–179). Houndmills: Palgrave Macmillan.

Checkel, J. T. (2008). Process Tracing. In A. Klotz and D. Prakesh (eds), *Qualitative Methods in International Relations: a Pluralist Guide* (pp. 114–127). New York: Palgrave Macmillan.

Dupont, C. and Oberthür, S. (2012). Insufficient Climate Policy Integration in EU Energy Policy: the Importance of the Long-Term Perspective. *Journal of Contemporary European Research*, 8(2), 228–247.

Dupont, C. and Oberthür, S. (2015). Conclusions: Lessons Learned. In C. Dupont and S. Oberthür (eds), *Decarbonization in the European Union: Internal Policies and External Strategies* (pp. 244–265). Houndmills: Palgrave Macmillan.

Edmondson, B. and Levy, S. (2013). *Climate Change and Order. The End of Prosperity and Democracy*. Houndmills: Palgrave Macmillan.

Eising, R. and Lehringer, S. (2010). Interest Groups and the European Union. In M. Cini and N. Pérez-Solórzano Borragán (eds), *European Union Politics* (pp. 189–206). Oxford: Oxford University Press.

European Commission. (2009). Adapting to Climate Change: Towards a European Framework for Action. *COM(2009) 147*.

European Commission. (2013). An EU Strategy on Adaptation to Climate Change. *COM(2013) 216*.

European Commission. (2014). Communication from the Commission to the European Parliament and the Council. Energy Efficiency and its Contribution to Energy Security and the 2030 Framework for Climate and Energy Policy. *COM(2014) 520*.

European Commission. (2015). Energy Union Package: a Framework Strategy for a Resilient Energy Union with Forward-Looking Climate Change Policy. COM(2015) 80.

Fioretos, O. (2011). Historical Institutionalism and International Relations. *International Organization*, 65(2), 367–399.

Gerring, J. (2008). Case Selection for Case-Analysis: Qualitative and Quantitative Techniques. In J. M. Box-Steffensmeier, H. E. Brady and D. Collier (eds), *The Oxford Handbook of Political Methodology* (pp. 645–684). Oxford: Oxford University Press.

Gullberg, A. T. (2008a). Lobbying Friends and Foes in Climate Policy: The Case of Business and Environmental Interest Groups in the European Union. *Energy Policy*, 36(8), 2964–2972.

Gullberg, A. T. (2008b). Rational Lobbying and EU Climate Policy. *International Environmental Agreements: Politics, Law and Economics*, 8(2), 161–178.

Haas, E. B. (1958). *The Uniting of Europe: Political, Social and Economic Forces, 1950–1957*. Stanford, CA: Stanford University Press.

Hauser, H. (2011). European Union Lobbying Post Lisbon: an Economic Analysis. *Berkeley Journal of International Law*, 29(2), 680–709.

Herodes, M., Adelle, C. and Pallemaerts, M. (2007). *Environmental Policy Integration at the EU Level: a Literature Review*. EPIGOV Paper No. 5. Berlin: Ecologic, Institute for International and European Environmental Policy.

Hopf, C. (2004). Qualitative Interviews: an Overview. In U. Flick, E. von Kardoff and I. Steinke (eds), *A Companion to Qualitative Research* (pp. 203–208). London: Sage.

Hovi, J., Sprinz, D. F. and Underdal, A. (2009). Implementing Long-Term Climate Policy: Time Inconsistency, Domestic Politics, International Anarchy. *Global Environmental Politics*, 9(3), 20–39.

IPCC. (2007). *Climate Change 2007. Fourth Assessment Report: Synthesis Report*. Geneva: Intergovernmental Panel on Climate Change.

IPCC. (2013). Summary for Policymakers. In T. F. Stoker, D. Qin, G.-K. Plattner, M. Tignor, S. K. Allen, J. Boschung, ... P. M. Midgley (eds), *Climate Change 2013: the Physical Science Basis. Contribution of Working Group I to the Fifth Assessment Report of the Intergovernmental Panel on Climate Change*. Cambridge: Cambridge University Press.

Jordan, A., Huitema, D., van Asselt, H., Rayner, T. and Berkhout, F. (2010). Governing Climate Change in the European Union: Understanding the Past and Preparing for the Future. In A. Jordan, D. Huitema, H. van Asselt, T. Rayner and F. Berkhout (eds), *Climate Change Policy in the European Union: Confronting the Dilemmas of Mitigation and Adaptation?* (pp. 253–275). Cambridge: Cambridge University Press.

Jordan, A. and Lenschow, A. (eds) (2008). *Innovation in Environmental Policy? Integrating the Environment for Sustainability*. Cheltenham: Edward Elgar.

Jordan, A. and Lenschow, A. (2010). Environmental Policy Integration: a State of the Art Review. *Environmental Policy and Governance*, 20(3), 147–158.

Kivimaa, P. and Mickwitz, P. (2006). The Challenge of Greening Technologies: Environmental Policy Integration in Finnish Technology Policies. *Research Policy*, 35, 729–744.

Kivimaa, P. and Mickwitz, P. (2009). *Making the Climate Count: Climate Policy Integration and Coherence in Finland. The Finnish Environment* (Vol. 3). Helsinki.

Klotz, A. (2008). Case Selection. In A. Klotz and D. Prakesh (eds), *Qualitative Methods in International Relations: a Pluralist Guide* (pp. 43–58). New York: Palgrave Macmillan.

Knudsen, J. (2012). Renewable Energy and Environmental Policy Integration: Renewable Fuel for the European Energy Policy? In F. Morata and I. Solorio Sandoval (eds),

European Energy Policy: an Environmental Approach (pp. 48–65). Cheltenham: Edward Elgar.

Lafferty, W. M. and Hovden, E. (2003). Environmental Policy Integration: Towards an Analytical Framework. *Environmental Politics*, 12(5), 1–22.

Lenschow, A. (ed.) (2002). *Environmental Policy Integration: Greening Sectoral Policies in Europe*. London: Earthscan.

Liberatore, A. (1997). The Integration of Sustainable Development Objectives into EU Policymaking. In S. Baker, M. Kousis, D. Richardson and S. Young (eds), *The Politics of Sustainable Development* (pp. 107–126). London: Routledge.

Michaelowa, A. (1998). Impact of Interest Groups on EU Climate Policy. *European Environment*, 8, 152–160.

Mickwitz, P., Aix, F., Beck, S., Carss, D., Ferrand, N., Görg, C., ... van Bommel, S. (2009). *Climate Policy Integration, Coherence and Governance. PEER Report No. 2*. Helsinki: Partnership for European Environmental Research.

Monaghan, E. (2013). Making the Environment Present: Political Representation, Democracy and Civil Society Organisations in EU Climate Change Politics. *Journal of European Integration*, 35(5), 601–618.

Moravcsik, A. and Schimmelfennig, F. (2009). Liberal Intergovernmentalism. In A. Wiener and T. Diez (eds), *European Integration Theory*, 2nd edn. (pp. 67–87). Oxford: Oxford University Press.

Nilsson, M. and Persson, Å. (2003). Framework for Analysing Environmental Policy Integration. *Journal of Environmental Policy and Planning*, 5(4), 333–359.

Nollkamper, A. (2002). Three Conceptions of the Integration Principle in International Environmental Law. In A. Lenschow (ed.), *Environmental Policy Integration: Greening Sectoral Policies in Europe* (pp. 22–34). London: Earthscan.

Oberthür, S. and Roche Kelly, C. (2008). EU Leadership in International Climate Policy: Achievements and Challenges. *International Spectator*, 43(3), 35–50.

Persson, Å. (2004). *Environmental Policy Integration: An Introduction. PINTS – Policy Integration for Sustainability Background Paper*. Stockholm: Stockholm Environment Institute.

Persson, Å. (2007). Different Perspectives on EPI. In M. Nilsson and K. Eckerberg (eds), *Environmental Policy Integration in Practice: Shaping Institutions for Learning* (pp. 25–48). London: Earthscan.

Pollack, M. A. (2009). The New Institutionalisms and European Integration. In A. Wiener and T. Diez (eds), *European Integration Theory*, 2nd edn. (pp. 125–143). Oxford: Oxford University Press.

Rietig, K. (2012). Climate Policy Integration Beyond Principled Priority: a Framework for Analysis. *Centre for Climate Change Economics and Policy, Working Paper No 99*.

Strøby-Jensen, C. (2007). Neo-functionalism. In M. Cini (ed.), *European Union Politics*, 2nd edn. (pp. 85–98). Oxford: Oxford University Press.

Tanasescu, I. (2009). *The European Commission and Interest Groups: Towards a Deliberative Interpretation of Stakeholder Involvement in EU Policy-Making*. Brussels: VUB Press.

Torney, D. (2014). External Perceptions and EU Foreign Policy Effectiveness: The Case of Climate Change. *JCMS: Journal of Common Market Studies* 52(6), 1358–1373.

Voss, J.-P., Smith, A. and Grin, J. (2009). Designing Long-Term Policy: Rethinking Transition Management. *Policy Sciences*, 42, 275–302.

Wurzel, R. K. W. and Connelly, J. (eds) (2011). *The European Union as a Leader in International Climate Change Politics*. London: Routledge.

List of interviews

No.	Interviewee	Date
1.	European renewable industry representative	26 January 2010
2.	DG Climate Action official	06 July 2010
3.	DG Climate Action official	06 July 2010
4.	Member of the Scottish Parliament	03 August 2010
5.	European climate NGO representative	10 January 2012
6.	European environmental NGO representative	09 February 2012
7.	European environmental NGO representative	15 February 2012
8.	European environmental NGO representative	7 May 2012 and 20 March 2013
9.	DG Energy official	25 June 2012
10.	European energy efficiency industry representative	05 July 2012
11.	European energy efficiency industry representative	12 July 2012
12.	DG Energy official	01 August 2012
13.	Gas industry representative	10 September 2012
14.	Estonian environmental NGO representative	14 September 2012
15.	Estonian environmental NGO representative	17 September 2012
16.	Estonian environmental NGO representative	18 September 2012
17.	Austrian environmental NGO representative	18 September 2012
18.	Latvian environmental NGO representative	18 and 21 September 2012
19.	DG Energy official	03 October 2012
20.	European climate research institute representative	24 October 2012
21.	Energy market analyst	23 January 2013
22.	Former DG Energy official	06 February 2013
23.	DG Climate Action official	15 March 2013
24.	DG Energy official	20 March 2013
25.	Member of the European Parliament	27 March 2013
26.	Member of the European Parliament	09 April 2013
27.	Energy consultant	04 November 2014
28.	Former DG Energy official	06 March 2015

Index

Page numbers in *italics* denote an illustration, **bold** indicates a table

ALTENER 64

biofuels: sustainability criteria debate 71, 72, 83; transport applications target 64, 71
Blair, Tony 19, 140
Briassoulis, Helen 32
Bryner, Gary 35
building industry associations: energy efficiency lobbying 109, 117–18
Buildings Performance Institute Europe (BPIE) 99–101

carbon capture and storage (CCS): climate and energy package (2008) 23, 71, 72; decarbonisation targets, role in 65, 127, 132
climate change: global problem 1; regional impact within Europe 13–14
climate policy integration (CPI): case study results 152–4, **153**; consultation access issues 161; definition 2; European integration theories, contributory factors 42–8; explanatory variables, contributory sequence 168–71, *169*; explanatory variables for framework 41, 48–54, 176–7, 179, 183; functional interrelations 37–8, 48, 50–1, **51**; functional interrelations of case studies 154–7, **157**, 170; future research projects 183–4; institutional and policy context 49–50, 51–2, 161–6, *166*; long-term policymaking challenge 181–2, 184; measuring in policy output 36, 38–40, **40**; measuring in policy process 36, 38, **39**; policy goals for 2050, analysis resources 40; political commitment 49, 51, **52**, 157–61, **161**; principled priority standard 35; priority setting of objectives 182; procedural changes across EU 182–3; process dimension (output level) 52–3, 166–8; pro-climate stakeholder involvement 36–7, 170–1; research methodology 4–7, 174–5; research process and analysis 175–8
CO_2/energy tax 18
Committee of Permanent Representatives (COREPER) 9–10
Committee of the Regions (COR) 11
Council of the EU: EEPR Regulation 2009 141; energy performance of buildings regulation and CPI 108, 115, 117–18, 119, 167; energy security strategies 97; EPBD 2002, implementation issues 105, 159; EPBD recast 2010, standards compliance 106–7, 119; political commitment to CPI 158–9, 160; RE Directive (2009), policy process 72–3; renewable energy policy, CPI levels 74–5, 76, 82, 84, 158–9, 167; RES-E Directive (2001), policy process 69–70; role within EU 9–10, 171, 180; TEN-E guidelines and revisions 138, 140

decarbonisation, 2050 targets: energy performance of buildings, low feasibility 101–2; long-term policymaking challenge 181–2; natural gas consumption estimates 131–2, *134*; natural gas, flawed potential 127, 144, 147, 156; renewable energy's potential role 65

DG Climate Action 36
DG Energy: energy performance of buildings and CPI 103–4, 107, 114; natural gas import infrastructure and CPI 145; renewables promotion and CPI 73, 81
DG Environment: assessment of CPI 36; climate and energy package (2008) 81; energy performance of buildings and CPI 107, 114; renewable energy and CPI 64, 73

elite socialisation 44, 48
emissions trading system (ETS) 22, 71, 72, 162
energy efficiency (EU economies) *see also* energy performance of buildings; Commission proposals (1970s-90s) 97–9; definition 95; energy consumption levels 95, 96, 97; energy intensity levels 95, 96; legislation **98**, 103; measurement and rebound effect 97
energy performance of buildings: 2050 targets and high CPI feasibility 101–3, *102*; climate policy objectives evaluated 110–11, 118–19, 122–3; Commission proposals (1970s-90s) 97–8; CPI, crucial explanatory variables 170; Directive 2002/91/EC, amendment manipulation 104–5, 162–3; Directive 2002/91/EC, CPI levels 107–14, *112*, *113*; Directive 2002/91/EC, proposals 103–5, 159; Directive 2010/31/EU, CPI levels 114–19; Directive 2010/31/EU, proposals 105–7; EU building sector study 99; existing buildings, renovations challenge 99–101; functional interrelations of CPI 110–11, 118–19, 155–6; institutional and policy factors effecting CPI 162–6, **166**; legislative improvements 103; national standards improved 97; nearly-zero and zero energy buildings 106–7, 119; political commitment to CPI 159–60, **161**; subsidiarity issue, effect on CPI 98, 107, 108, 167
Energy Performance of Buildings Directive (EPBD) 2002: external stakeholders and CPI levels 108–10; functional interrelations recognised 110–11; internal stakeholders and CPI levels 107–8; policy output and CPI level 111–14, *112*, *113*; proposals and debate 103–5
Energy Performance of Buildings recast Directive (EPBD recast) 2010: external stakeholders and CPI levels 116–18; functional interrelations recognised 118–19; internal stakeholders and CPI levels 114–16; policy output and CPI level 119–21, *120–1*; proposal and debate 105–7
energy policy (EU): 'catch-up governance' 180; climate and energy package (2008) 23, 71, 72–3, 81, 164, 179; climate policy objectives evaluated 88–9, 121–3, 148, 152–4, **153**, 176–8; Commission proposals 9, 14, 19, 21–2, 23; consultation access issues 161, 180; Council of the EU's role 10, 171, 180; EEPR Regulation 2009, policy process 140–2; energy efficiencies, 'market failure' 101, 103; energy efficiency, Commission proposals (1970s-90s) 97–9; energy efficiency legislation **98**, 103, 122; energy infrastructure development 129–30; energy performance of buildings, improvement challenge 99–101, 122–3; energy security strategies 63, 97, 130–1, 135–7, 148, 165; EPBD 2002, proposals and debate 103–5; EPBD Recast 2010, proposals and debate 105–7; European Council, major influence 11, 16, 18, 22–3; external stakeholder contributors 11–12; institutional contexts and CPI 163–4, 179; internal energy market 14, 16, 18, 19, 63–4, 135–6; international agreements, impact on 16, 18–19, 22–3; international climate negotiation schedules 164–5; main policy developments 14–23, **15**, **17**, **20**; membership enlargement and negotiation impacts 161–2; natural gas consumption and decarbonisation targets 2050 131–2; natural gas import infrastructure development 128–9, **129**, 132–3; natural gas import infrastructure, financial support 140–2, 144–5; Parliamentary role limited 10–11; path dependency limitations 163; political commitment 157–8; RE Directive (2009), proposals and debate 70–3; renewable energy, 2050

Index   191

objectives 64–6, 66; renewable energy, early developments 62–4; renewable energy roadmap 70; RES-E Directive (2001), proposals and debate 67–70; support schemes for renewables 68–9, 73, 76; TEN-E guidelines and revisions 137–40

ENVI Committee (EU Parliament): assessment of CPI 36; energy performance of buildings, CPI levels 107–8, 115; renewable energy policy, CPI levels 73–4, 82; RES-E Directive (2001), policy process 69, 71; TEN-E guidelines and revisions 139, 142

Environmental Action Programme (EAP) 33

environmental NGOs: energy efficiency (2002), limited response 108–9; energy efficiency (2010), accessible consultation 116–17; gas import infrastructure, limited response 143, 146; renewable energy lobbying 75–7, 83–4

environmental policy integration (EPI): concept interpretation 34; definition 2; development within EU policymaking 33–4; explanatory variables 41; normative standards 34–5; relevance in policymaking 183–4

Eurogas 131, 133

European Climate Change Programme (ECCP) 21–2, 109

European Commission (EC): climate and energy package (2008) 23, 71, 81; climate change proposals 16, 18, 21–2; climate policy and energy related proposals 68, 99, 105–6, 118, 159; CO2/energy tax 18; DG Energy and CPI 73, 81, 103–4, 107, 114, 145; DG Environment and CPI 64, 73, 107, 114; EEPR Regulation 2009 145–6; energy efficiency proposals (1970s–90s) 97–9; energy performance of buildings and CPI 103–4, 107, 110–11, 114, 118, 159–60; energy policy, first guidelines 14; energy roadmap to 2050 132; energy security strategies 19, 21, 130, 135–6; Energy Union plans 148; environmental NGOs involvement 75–6, 83; European Energy Programme for Recovery (EEPR) Regulation 2009 141–2; policymaking process 9; RE Directive (2009), proposals and debate 71, 73; renewable energy, EU-level action 64; renewable energy projections 66, 66; renewables and climate change objectives 85, 158–9; RES-E Directive (2001), proposals and debate 67–70; trans-European networks for energy (TEN-E) 137–40

European Council: climate change, action sort 16, 18; emissions reduction targets 22–3, 158; energy security strategies 19; policymaking, major impetus 11; renewables and climate change objectives 84–5

European Economic and Social Committee (EESC) 11

European Energy Programme for Recovery (EEPR) Regulation 2009: policy output and CPI 147–8; policy process 140–2; policy process and CPI 145–7

European integration theories: contribution to CPI framework 42–3; institutionalist-centred theories 46–8; liberal intergovernmentalism 44–6; neofunctionalism and integration 43–4

European Parliament: climate change resolutions 16; EEPR Regulation 2009 141, 164; energy policy, limited influence 10–11; ENVI Committee and CPI 36, 82, 107–8, 115, 142; environmental NGOs involvement 76–7, 83–4; EPBD 2002, CPI suggestions dismissed 107–8, 159; EPBD 2002, financing issues 104; EPBD recast 2010, financial incentives 106; EPBD recast 2010, pro-climate MEPs 114–15, 118–19; political commitment to CPI 158, 159, 160; RE Directive (2009), proposals and debate 71–3, 82; RES-E Directive (2001) proposals 68–70; TEN-E guidelines and revisions 138, 139–40, 142

European Renewable Energy Council (EREC) 76

European Union (EU): climate change, leadership strategy 1, 2, 7, 23, 157–8; climate policy, main developments 14, **15**, 16, **17**, 18, **20**, 21–3; Commission's role 9; Council of EU's role 9–10; emissions trading system (ETS) 22; energy policy and greenhouse gas emissions 3–4, 64; energy policy, main developments 14, **15**, 16, **17**, 18–19, **20**; Kyoto negotiations 18–19;

192   Index

legislative procedure 12–13, 161–2; Parliament's role 10–11; policymaking process 7–11; regional impact of climate change 13–14; sustainable development 33
explanatory variables, CPI framework: combined strengths 53–4; functional interrelations 50–1, **51**; institutional and policy context 51–2; political commitment 51, **52**; process dimension (output level) 52–3; summary **55**

functional interrelations: case studies analysed 154–7, **157**, 170; energy performance of buildings 110–11, 118–19, 155–6; explanatory variable 48, 50–1; natural gas import infrastructure 143–4, 146–7; renewable energy 64–86, 77–8, 155

greenhouse gas emissions (GHG): EU reduction objectives 3–4, 22–3, 64, 85, 100; main sources 13

Haas, Ernst 43
Hovden, Eivind 34

institutionalist-centred theories 46–8, 49–50
Intergovernmental Panel on Climate Change (IPCC) 1, 16
ITRE Committee (EU Parliament): RE Directive (2009) debate 72; renewable energy policy, CPI levels 73–4, 82; RES-E Directive (2001) debate 69–70, 71

Kyoto Protocol 19, 22

Lafferty, William M. 34
liberal intergovernmentalism 44–6, 49

member states: CPI integration 48, 49; energy performance of buildings implementation 105, 106–7, 119, 162–3; membership enlargement and negotiation impacts 162–3; motivational factors (integration theories) 42–3, 45–6, 47; renewable energy, consumption share 61–2, **62**; renewables and electricity supply targets 68–70; representation with EU 9–11
Moravcsik, Andrew 44–6

natural gas import infrastructure: capacity expansion and investment 128–9, **129**, 132–3; carbon capture and storage (CCS) 127, 132; climate change objectives, conflict sidelined 143–4, 146–7, 156, 167–8, 170; Connecting Europe Facility (CEF) 148; CPI, crucial explanatory variables 170; Directive 2004/67/EC 135–6; energy consumption and supply levels 128; energy security strategies 130–1, 135–7, 161, 165; EU energy infrastructure policy 129–30; EU promotional policies of 2000s 135, 163; European Energy Programme for Recovery (EEPR) Regulation 2009 140–2, 145–8; extraction and transportation methods 127–8; financial support 140–2, 144–5; functional interrelations of CPI 143–4, 146–7; institutional and policy factors effecting CPI 162–6, **166**; natural gas consumption and decarbonisation targets 2050 131–3, *134*; political commitment to CPI 160–1, **161**; 'project of European interest' proposal 139–40, 144–5; trans-European networks for energy (TEN-E) 137–40, 142–5
nearly-zero-energy buildings: energy efficiency renovations 100, 102; regulatory negotiations 106–7, 119
neofunctionalism and integration 43–4, 48
nuclear energy 65

policy coherence 30–2
policy coordination 30–2
policy integration: key concepts 30–2; variances in conceptualisation 32
policy process and CPI: climate policy objective incorporation 37–8; measuring CPI 36, **39**; pro-climate stakeholder involvement 36–7
policy process (integration theories): 'Europeanised' policy makers 44; neofunctionalist 'spillover' concepts 43–4; process-centred, institutionalist view 46–8; state-centred and rationalist approach 44–5; supranational interest groups 44
political commitment (CPI variable) 49, 51, **52**, 157–61, **161**
process tracing techniques 6–7
pro-climate policy stakeholders: access for

actors 36–7; external actors and CPI levels 75–7, 83–4, 108–10, 116–18, 143; internal actors and CPI levels 73–5, 81–2, 107–8, 114–16, 142

qualified majority voting rule (QMV) 9

rebound effect 97
RE Directive (2009): external stakeholders and CPI levels 83–4; functional interrelations recognised 84–6; internal stakeholders and CPI levels 81–2; policy output and CPI level 86–7, 87, 88; proposals and debate 70–3
renewable energy, EU policy: climate and energy package (2008) 23, 71, 72–3, 164; climate policy objectives evaluated 88–9, 153, 155; consultation access issues 161, 166–7; consumption share increases 61–2, 62, **63**; CPI, benchmarks and projection 64–6, 66; CPI, crucial explanatory variables 169–70; defined sources 61; early developments 62–4; electricity supply targets 68–70; energy infrastructure development 130; functional interrelations of CPI 77–8, 84–6, 155, **157**; institutional and policy factors effecting CPI 162–6, **166**; political commitment to CPI 158, **161**; promotion through RES-E Directive (2001) 67–8, 77–8; RE Directive (2009), policy output and CPI 86–7, 87, 88; RE Directive (2009), policy process and CPI 81–6; RE Directive (2009), proposals and debate 70–3; renewables trading debate 71, 72; RES-E Directive (2001), policy output and CPI 78–81, 79, 80; RES-E Directive (2001), policy process and CPI 73–8; RES-E Directive (2001), proposals and debate 67–70; roadmap objectives 70; support schemes 68–9, 73, 76
renewables in transport 71
renewables trading 71
research methodology: case study selection 4–6, 175; data sources 6–7; explanatory variables framework 176–7, 179; objectives of research 174–5; operationalisation of CPI 175–6; scholarly contribution 178–80
RES-E Directive (2001): external stakeholders and CPI level 75–7; functional interrelations recognised 77–8; internal stakeholders and CPI level 73–5; policy output and CPI level 78–81, 79, 80; proposals and debate 67–70
Rothe, Mechtild 68–9, 74

SAVE programme 103
Single European Act 1987 16, 33
spillover concepts: explanatory variables, CPI framework 48–9; theoretical basis 43–4
supranational interest groups 44, 48

trans-European networks for energy (TEN-E): external stakeholders and CPI levels 143; functional interrelations, lack of 143–4; internal stakeholders and CPI levels 142; policy output and CPI level 144–5; policy process and debate 137–40
Treaty on the Functioning of the European Union (TFEU): energy infrastructure development 129–30; energy policy limitations 21; environmental protection 2, 33
Turmes, Claude 71, 82, 83–4, 115

Underdal, Arild 32
UN Environment Programme (UNEP) 16
UN Framework Convention on Climate Change (UNFCCC) 7, 13, 18

Wijkman, Anders 71, 82
World Meteorological Organisation (WMO) 16

zero energy buildings: new build standards 101; regulatory negotiations 106–7

eBooks
from Taylor & Francis

Helping you to choose the right eBooks for your Library

Add to your library's digital collection today with Taylor & Francis eBooks. We have over 50,000 eBooks in the Humanities, Social Sciences, Behavioural Sciences, Built Environment and Law, from leading imprints, including Routledge, Focal Press and Psychology Press.

Choose from a range of subject packages or create your own!

Benefits for you
- Free MARC records
- COUNTER-compliant usage statistics
- Flexible purchase and pricing options
- All titles DRM-free.

Benefits for your user
- Off-site, anytime access via Athens or referring URL
- Print or copy pages or chapters
- Full content search
- Bookmark, highlight and annotate text
- Access to thousands of pages of quality research at the click of a button.

Free Trials Available
We offer free trials to qualifying academic, corporate and government customers.

eCollections

Choose from over 30 subject eCollections, including:

Archaeology	Language Learning
Architecture	Law
Asian Studies	Literature
Business & Management	Media & Communication
Classical Studies	Middle East Studies
Construction	Music
Creative & Media Arts	Philosophy
Criminology & Criminal Justice	Planning
Economics	Politics
Education	Psychology & Mental Health
Energy	Religion
Engineering	Security
English Language & Linguistics	Social Work
Environment & Sustainability	Sociology
Geography	Sport
Health Studies	Theatre & Performance
History	Tourism, Hospitality & Events

For more information, pricing enquiries or to order a free trial, please contact your local sales team:
www.tandfebooks.com/page/sales

www.tandfebooks.com